Smart University

CRITICAL UNIVERSITY STUDIES
Jeffrey Williams and Christopher Newfield, Series Editors

Smart University

Student Surveillance in the Digital Age

Lindsay Weinberg

JOHNS HOPKINS UNIVERSITY PRESS   BALTIMORE

© 2024 Johns Hopkins University Press
All rights reserved. Published 2024
Printed in the United States of America on acid-free paper
9 8 7 6 5 4 3 2 1

Johns Hopkins University Press
2715 North Charles Street
Baltimore, Maryland 21218
www.press.jhu.edu

Library of Congress Cataloging-in-Publication Data

Names: Weinberg, Lindsay, 1990– author.
Title: Smart university : student surveillance in the digital age / Lindsay Weinberg.
Description: Baltimore : Johns Hopkins University Press, [2024] | Series: Critical university studies | Includes bibliographical references and index.
Identifiers: LCCN 2024010347 | ISBN 9781421450018 (hardcover ; acid-free paper) | ISBN 9781421450025 (ebook)
Subjects: LCSH: Education, Higher—Effect of technological innovations on. | Electronic surveillance—Social aspects. | Privacy, Right of.
Classification: LCC LB2395.7 .W456 2024 | DDC 378.00285—dc23/eng/20240513
LC record available at https://lccn.loc.gov/2024010347

A catalog record for this book is available from the British Library.

Special discounts are available for bulk purchases of this book. For more information, please contact Special Sales at specialsales@jh.edu.

Contents

Acknowledgments vii

Introduction 1

1 The "Smart" University 22

2 Recruitment 48

3 Retention 65

4 Wellness 83

5 Security 111

6 (Beyond) Digital Literacy 150

Conclusion. Against the Smart University 162

Notes 171
Bibliography 205
Index 233

Acknowledgments

When I began a postdoctoral fellowship at Purdue University in 2018, I learned that many of the digital marketing techniques I had researched while a graduate student in the History of Consciousness Department at the University of California, Santa Cruz, were being pursued as solutions to seemingly intractable problems in US higher education. This book is indebted to the guidance I received from Robert Meister, Carla Freccero, Warren Sack, and Mark Andrejevic on that dissertation research, as it significantly influenced how I've come to understand the dynamics of industry power, surveillance, and injustice animating the rise of the smart university.

I have also received tremendous support from colleagues in the John Martinson Honors College at Purdue, especially Megha Anwer, Anish Vanaik, Nathan Swanson, Natasha Duncan, Kristina Bross, and Emily Allen. Conversation and collaborations with Megha have significantly shaped my understanding of the politics of diversity within higher education. Her dear friendship, and the friendships of Christina Neri, Daisy Griggs, Rashinda Reed, Antonella Raczynski, Lara Galas, Stephen David Engel, Lindsay Keebler, and Caitlin McNichol, have sustained me in countless ways throughout the writing process. I am also grateful to Megha, Scott Henkel, Surya Parekh, Ken Roark, and Chris Clifton for their mentorship over many years. My thinking about resistance within the smart university is indebted to Scott's feedback and transformative teaching. The love and constant encouragement I received from Steve and Josh Weinberg, Elly and Eli Cohn, and Muiris MacGiollabhuí made this project possible.

This book's argument has also been honed through collaborations with students, staff, and faculty in the Critical Data Studies Collective and the Tech Justice Lab at Purdue. In particular, student researchers Gary Chen, Nikita Gerard, and Antonio Dominguez Palomar contributed to early literature review efforts that shaped this book's first two chapters. Conversations with participants in the Critical Data Studies Collective's 2021 teach-in "Do Black Lives Matter to the University? Science, Technology, and the Racial-Colonial Legacies of Higher Education," featuring Davarian Baldwin, Darrin Johnson, and Audrey Beard, and co-organized with Marlo David, Danielle Walker, Risa Cromer, and Laura Zanotti, have also informed how I situate contemporary campus security measures within longer university histories of policing. Madisson Whitman offered crucial feedback on the book manuscript both conceptually and organizationally, which helped me think about how to convey the significance of the smart university in a context where digital technologies are diffused throughout many facets of academic life. I have also been fortunate to share part of this work with students and faculty through talks at the University of California, Irvine, and the Surveillance Studies Network Conference. Pointed questions from audience members helped me to improve this book's chapters on retention and wellness.

During the final stages of writing, I was fortunate to receive summer funding from Purdue University's Office of Research, as well as a fellowship with the Aarhus Institute of Advanced Studies (AIAS) and the Shaping Digital Citizenship research center at Aarhus University. AIAS provided a generative interdisciplinary community for feedback from other fellows as well as colleagues affiliated with the Centers for Science Studies and Science and Technology Studies. This feedback helped me further consider the book's implications for students and faculty outside the United States. Jeremy Brown offered formatting assistance that I am particularly grateful for. I am also indebted to Hopkins Press editor Greg Britton's encouraging feedback and guidance throughout multiple stages of the project, to the instructive comments of anonymous peer reviewers, and to Carrie Watterson's diligent copyediting.

This book is for my mother, Nina Weinberg, and for every student I've had the privilege of teaching and learning from. Discussions with students about the relationship between digital technology, higher education, and student agency have inspired me throughout the writing of this book. These conversations and the wave of recent student debt resistance, labor organizing, and abolitionist struggle sweeping higher education affirm the power of coalitions to transform universities for the better. I hope this book honors these collective efforts to reimagine higher education and redress its longstanding harms.

Part of chapter 4 was first published as "Mental Health and the Self-Tracking Student" in *Catalyst: Feminism, Theory, Technoscience* 7, no. 1 (2021): 1–27. Part of chapter 5 first appeared in the article "Feminist Research Ethics and Student Privacy in the Age of AI," which was published in *Catalyst: Feminism, Theory, Technoscience* 6, no. 2 (2020): 1–10.

Smart University

Introduction

Higher education is becoming increasingly synonymous with digital surveillance in the United States. Advanced network infrastructure, internet-connected devices and sensors, radio frequency identification (RFID), data analytics, and artificial intelligence (AI) are being celebrated as a means of ushering in the age of "smart universities," one where institutions can run their services more efficiently and strengthen the quality of higher education using digital tools.[1] However, as this book demonstrates, these tools have a darker side. They allow public universities to respond to and perpetuate corporate logics of austerity, use student data to reduce risk of financial investment in the face of dwindling public resources, and track student behavior to encourage compliance with institutional metrics of success. Surveillance of student behavior forms the foundation of the smart university, often in ways that prove harmful to students—particularly those who are already marginalized within the academy.

Smart university initiatives generally include the use of data-intensive digital technologies to do some or all of the following: monitor and automate aspects of student learning, extracurricular participation, and progress to degree; manage facilities and resources; produce new revenue streams; support research activities that lead to external funding; and purportedly enhance campus security and student wellness. The idea of smart universities builds on popular visions and practical applications of "smart" technology—originally, an acronym for "self-monitoring, analysis, and reporting technology"—that has since become associated with a range of devices that incorporate automation, internet connectivity, data analytics, or intuitive user interfaces.[2] According to proponents like vendors of smart university solutions, higher education administrators, and campus information technology professionals, these technologies will theoretically reduce costs and improve campus sustainability.[3] Furthermore, while some of these initiatives may not explicitly use the language of smartness (preferring instead "digital campus," "intelligent campus," or other technocentric terms), they nonetheless rely on digital surveillance deployed in the name of improving campus inefficiencies or the student learning experience.

This book uses the term "smart university" broadly to illustrate how digital transformations in public higher education are being embedded with shared austerity logics across many aspects of student life. It considers how these recent changes offer lenses into the wider social and economic problems shaping inequity in higher education. It's important to note early and often that smart universities are not merely a suite of technologies being implemented on college campuses. These initiatives emerge from and enact visions of what is most necessary for the future of higher education, or what science and technology studies scholars Sheila Jasanoff and Sang-Hyun Kim refer to as sociotechnical imaginaries: "collectively held, institutionally stabilized, and publicly performed visions of desirable futures."[4] More generally, smart technologies are bound up with ideas of how to promote resilience in the wake of uncertainty, financial crisis, and environmental degradation.[5] The increased penetration of smart technologies into everyday life will theoretically allow for these crises to be perpetually managed. This book treads carefully, documenting how the proliferation of digital technologies for making universities "smarter" is indeed bound up with dramatic cuts to public higher education spending. At the same time, it demonstrates that present-day issues of digital surveillance and inequity in public universities come from deeply rooted, long-standing injustices woven into the fabric of US higher education's history.

Proponents of smart universities argue that public universities need to draw from the data analytics strategies of corporations like Amazon and Netflix to truly know their student body. This knowledge about students requires a wealth of data collection about prospective and current students' behaviors. Furthermore, the data is often acquired in partnership with for-profit firms with little regard for student privacy. Students' data can inform the recruitment messages they receive, the courses they are encouraged to take, and whether they are flagged for real-time advising or wellness interventions in the hopes of influencing student behavior. A range of public universities across the United States are being restructured around this production and capture of digital data.

Many industry insiders and technologists celebrate corporations' use of data analytics as a means of delivering more personalized and attractive consumer options. However, critics have pointed out that this type of personalized digital targeting can have exploitative and prejudicial consequences. There have been reports of digital targeted marketing leading to price discrimination against low-income users and to the circumventing of civil rights–era protections against job and housing discrimination.[6] Despite these concerns, predictive analytics—the use of data, statistics, and modeling techniques to make forecasts about the future—are increasingly attractive to US public higher education administrators in a funding-poor but data-rich education landscape. This book investigates the emergence and consequences of this proliferation of data analytics and related digital technologies in the smart university.

While the term "smart campus" was originally coined in 2000 and has since become the preferred industry buzzword,[7] the rise of these initiatives in the United States more broadly coincides with a set of shifts often associated with the neoliberal university, meaning universities structured according to capitalist market principles with increased managerial control, beginning in the 1980s. These transformations have included skyrocketing tuition, funding cuts to mental healthcare, increased corporate partnerships, rapid rises in massive open online courses (MOOCs), and the expansion of campus real estate both regionally and globally.[8]

Simultaneously, faculty have faced higher education administrators' increasing use of performance metrics, impact measures, citation counts, and other quantitative, market-informed indicators for decisions about employability and job security.[9] In addition to administrators, major proponents of the smart university include university researchers and information technology specialists, industry providers of smart infrastructure and education technology, think tanks, and consultancies. More often than not, and without supplying credible evidence, advocates argue that students are increasingly expecting smarter campuses with "integrated learning experiences and streamlined student services," and that such initiatives are necessary to boost enrollments and maximize operational effectiveness.[10]

In an effort to call attention to the relationship between higher education and student welfare in the age of increasingly smart universities, I also focus on students' abilities to exercise their rights, access opportunities, and participate in institutional change. Drawing from science and technology studies (STS) and critical university studies, the book builds on the work of education technology experts like Ben Williamson, Neil Selwyn, and Audrey Watters who have linked the emergence of digital education governance to neoliberal ideology and policy.[11] This book argues that these more recent transformations are best understood as part of a longer history in which one of the university's primary functions is to support dominant techno-economic paradigms that reproduce racial and economic injustice.

Situating the Smart University

The relationship between technology and higher education governance has long been a site of struggle in US universities. For instance, in 1964, as part of the Free Speech Movement, University of California, Berkeley, protestors used punch cards as a metaphor for stifling university bureaucracy and alienation. By folding, spindling, and mutilating these punch cards, university students were not simply defying the printed warnings that appeared on punch cards at the time: they were rejecting the increasingly impersonal university space, restrictions on campus political activities, and the automation of the university.[12] This critique is perhaps best captured in Mario Savio's sit-in address on the steps of Sproul Hall in 1964:

> We have an autocracy which—which runs this university. It's managed . . . the faculty are a bunch of employees and we're the raw material! But we're a bunch of raw materials that don't mean to be—have any process upon us. Don't mean to be made into any product! Don't mean—don't mean to end up being bought by some clients of the university, be they the government, be they industry, be they organized labor, be they anyone! We're human beings![13]

Savio's speech indicted the university for training students to meet the demands of capitalism and US imperialism. The idea that students were raw material for the university was a metaphor for student dehumanization, including the denial of their rights and the suppression of their political will, within the education system.

This notion of the university as a bureaucratic machine used to produce obedience and conformity was a pervasive critique within the student movement. Berkeley students saw a clear link between emerging information technologies and the increasing depersonalization of the education system. Punch cards and "information machines" were frequently referenced in their flyers, pamphlets, and speeches. They were a marker of institutional control, from police departments to universities, which used punch cards for registration and recordkeeping. By the 1960s, punch cards, for many, represented a society where the demands of machines were being given priority over the needs and desires of human beings.[14]

Punch cards were certainly not the first use of recordkeeping for student enrollment and attendance. Schools in New England kept registers as early as the 1820s. However, advancements in modern communications technology and the political unrest of the late 1960s and early 1970s led to increased student monitoring in the United States. Researchers, educators, and social services collected growing amounts of student information during this period, including information about their families, values, sexuality, and drug use.[15] As student advocate Diane Divoky explained in 1973, the growing bureaucratization and centralization of schools had a snowball effect on information collection: "There was little thought given to development of clear policies and practices by which student and parental rights of privacy might be balanced against the needs of the school and other social agencies to know, or to guarantee, that material contained in records was accurate and pertinent."[16] Divoky was responding, in part, to a 1969 Russell Sage Foundation conference report. This report detailed a lack of informed consent and transparency, unregulated access by unauthorized school personnel, and an ad hoc approach to granting student data access to law enforcement officials, courts, employers, colleges, and researchers. Divoky was one voice among a growing chorus of parents demanding greater openness and transparency in educational recordkeeping processes, with increased distrust in the US government in the aftermath of the Watergate scandal. It was President Gerald Ford who signed the Family Educational Rights and Privacy Act (FERPA) into law on August 21, 1974, two weeks after he took office following Nixon's resignation. This act was designed to provide a right of access to educational records and to protect individual rights to privacy by "limiting the transferability of records without a student or parent's consent."[17] Senator James Buckley introduced the act in the hopes of remedying what the public saw as frequent school violations of student and parent privacy by providing these groups with greater knowledge and control over school information disclosure practices.

Recent times have not been without their fair share of political scandals regarding the security of personal information. For instance, I could point to Edward Snowden's 2013 revelations of sweeping government surveillance programs, where private telecommunications companies were handing over customers' data to the National Security Administration, or to the 2016 Cambridge Analytica scandal, where digital consultants on the Trump campaign exploited Facebook user data to build voter profiles and influence their behavior. Both cases illustrate the ways digital technologies are undermining privacy rights in the United States, rights that are already rarely protected for poor people, people of color, and immigrants.[18] Furthermore, social media platforms like Facebook (now Meta) have been sued for allowing advertisers to circumvent the Fair Housing Act and discriminate against users on the basis of race, gender, and other characteristics.[19] And yet, as this book demonstrates, Big Tech companies are now being used as models for innovating recruitment, retention, and student services in higher education. Why, and how, did this happen?

Public policy has largely moved away from addressing concerns over student data use practices, beginning with the passage of the Crime Awareness and Campus Security Act in 1990. During this period, policy shifted toward overriding privacy protections, especially in contexts involving campus safety and security. Throughout the 1990s, FERPA was amended to exclude campus law enforcement records from FERPA protections and provide the military access to student recruitment information. In the case of international students with F, J, and M visas, colleges and universities were required to report to the federal government personal information such as names, addresses, academic status, full-time attendance, and any disciplinary actions taken as a result of a crime. With the passage of the PATRIOT Act in 2001, institutions, including higher education, were made immune from liability for the good-faith release of records to an assistant attorney general or a higher-ranking federal office or employee in response to an ex parte order of the court. FERPA, meanwhile, was not amended to address student access and confidentiality in an increasingly digital age.[20]

At present, collected student data reserves are often put to use for secondary purposes, both educational and noneducational, and involve for-profit companies. FERPA not only fails to address student privacy concerns raised by big data analytics and data mining, but in fact it enables public-private data sharing. Universities have broad latitude in determining whether to share student data with private vendors, and third-party services are increasingly being used for recruitment, retention, and instructional purposes. Notice and consent requirements do little to shore up student privacy, given that many students need to use these services throughout their college experience. Additionally, whatever degree of transparency privacy policies may offer, students are rarely empowered to have control over, or change, the terms of these policies.

Schools are also permitted to share student data without consent to a "school official," who after a 2008 amendment to FERPA was defined to include contractors, consultants, volunteers, and other outsourced institutional service providers the school would otherwise use employees to perform. According to the amendment, there must be "legitimate educational interests" in the records as defined by the school. Universities have discretion over what counts as a "legitimate educational interest," and so this flexibility could permit schools to potentially sell student information for funding purposes.[21] Under conditions of austerity, where public funding for education is increasingly curtailed and restricted, student data is especially vulnerable to a wide range of uses with minimal constraints. Furthermore, not only does the practice of sharing student data with little accountability or oversight raise privacy issues, but these practices also permit student data to be exploited for the purposes of creating and improving private firms' products and services. In this sense, private firms are able to save money on what would otherwise require investment in market research and product development by virtue of being able to "put to work" the student data they collect.

Now certainly, there are ways digital tools can enhance student learning and help educators better understand students' learning processes. However, they are being introduced within a particular historical and institutional context. Technologies do not exist outside the world and then act upon it; rather, they are produced and embedded within complex (and often highly unequal) societies. Our tools contain implicit and explicit values, goals, judgments, and assumptions that shape the political horizon of what seems possible to transform about our social world, including higher education. For this reason, technologies are not neutral or apolitical.[22] In their most practical applications, we could also understand smart university technologies, drawing from the work of sociologists Pierre Lascoumes and Patrick Le Galès, as public policy instruments: tools for exercising social control that organize the higher education sector according to projected or calculated future risks, be they social, environmental, financial, or demographic.[23]

The primary set of social and historical circumstances most immediately shaping the introduction of these digital tools into the university is the set of governance strategies and political ideologies of neoliberalism, widely implemented in the 1980s in the United States under the Reagan administration. This period brought about a series of economic liberalization policies such as austerity, privatization, and deregulation. During his successful 1966 gubernatorial race in California, which set the stage for his political ascendency, Ronald Reagan had promised to use higher education budget cuts and the instantiation of tuition fees to "clean up the mess in Berkeley."[24] Reagan was referring to the growth of student civil rights and anti-war activism, including student civil disobedience associated with the Free Speech Movement. For the public university under neoliberalism, it became necessary to assume the characteristics of a business to respond to the new dominant socioeconomic paradigm. When state government support declined, publicly subsidized higher education came to be supplanted by a greater reliance on tuition payments, and deregulation allowed for the expansion of for-profit colleges. This debt-driven system coincided with shifting ideas about higher education as an individual investment rather than a social good. This shift, as sociologist Tressie McMillan Cottom explains,

> made the politics of financializing college tuition a sensible public choice. This attitude and the attendant policy changes proliferated: declining state investments in public higher education, austerity policies at community colleges and four-year colleges, incentives for universities to act more "entrepreneurial" and raise money. In effect, we told students that higher education was a private good with social rewards. Based on this, students should invest even as colleges increasingly shifted the costs of attendance to students. Investment implies risks and rewards.[25]

Smart universities play a role in this story because in many cases, universities see these initiatives as a means of acting more entrepreneurial, building partnerships with private firms, and taking on their characteristics and marketing strategies. These efforts are often imagined as vehicles for universities to counteract a lack of public funding sources and preserve their rankings in an education market students are increasingly priced out of. The prestige conferred by expensive elite institutions on the one hand and career-focused programs offered through more affordable community colleges and for-profit institutions on the other has intensified competition for recruiting and retaining students in the wake of unprecedented drops in enrollment. With steep rises in the cost of higher education and demands for greater accountability over student outcomes, there have been rising pressures from federal and state officials, accreditation agencies, major foundations and funders, governing boards, and media sources for universities to collect ever more fine-grained data about the institution and its students.

Selling the Smart University

Higher education administrators and digital technology providers often mobilize the rhetoric of techno-utopianism to sell smart university initiatives. Data analytics are frequently marketed as providing meaningful, personalized recruitment and learning experiences to promote positive reception, obscuring questions of data ethics. For instance, as administrators Jay W. Goff and Christopher M. Shaffer argue:

> In today's competitive enrollment environment, it is not enough to claim that your college is "student-centric" and knows your students' aggregate profile in terms of sociodemographics and entrance scores. Much like commercial organizations, universities must increasingly get to know their students, and the students whom they desire to enroll, as individuals—most students expect this level of familiarity. They are used to data clustering applications that help companies move quickly and proactively to meet their needs. Many students now expect all organizations, including higher education institutions, to respond in an almost hyper-personalized manner.[26]

Goff and Shaffer see a horizon of possibilities for universities if they emulate the practices of companies like eHarmony, Amazon, and Netflix, who have all used customer data to tailor their interactions with consumers. By truly "knowing" everything about a student's needs and abilities, Goff and Shaffer contend, universities can improve their recruitment and retention efforts. This knowledge about the student is predicated on the mass collection of student data, which for-profit software companies help universities compile and analyze. Prospective student data is used to segment the market and generate predictions of which prospective students are more likely to attend and the kinds of appeals and financial incentives they might find desirable during recruitment. Personalization is thus a technique of what political scientist Peter Miller and sociologist Nikolas S. Rose describe as neoliberal governmentality: a means of shaping the conduct of individuals in ways that allow for the future to be better managed.[27]

By controlling the options and messages students are confronted with, universities mobilize students' self-regulating and self-governing capacities—their ability to exercise choice—for compliance with institutional metrics of success under austerity. These efforts have disparately harmed low-income students and students of color. For reasons this book explains, these students' data doubles are less likely to fit the idealized student model baked into recruitment and retention initiatives, which often privilege structurally advantaged students. "Data doubles," here, refers to the outcome of using surveillance to monitor and abstract student behavior into flows of data that can be aggregated, mined, and then sorted and reassembled for targeted intervention.[28]

Smart university initiatives are, perhaps most significantly, a site of big data socialization, helping to produce students as data subjects capable of being managed and who, simultaneously, might go on to fill roles in the digital economy. For instance, at Virginia State University, partnerships with Microsoft, Abbott, and Raytheon Technologies are presented as a way train students to meet workforce needs resulting from the "complete integration of industry 4.0 or 'smart' technologies into so many aspects of our lives."[29] This partnership is part of a larger 2022 initiative between historically Black colleges and universities (HBCUs) and Fortune 100 corporate partners to diversify cybersecurity and defense fields. Additionally, in

2005, University of Memphis researchers partnered with IBM and the City of Memphis to develop Blue CRUSH (criminal reduction utilizing statistical history) for the purposes of predictive policing. This relationship between university interests, private industry, and state investments in the prison and military industrial complexes means that university research is directly implicated in creating and diffusing digital tools for the "permanent war and punishment economy," meaning the stitching together of capital, the state, and the academy in ways that further militarism and mass incarceration.[30]

While significant research partnerships between the university and militaristic national policy priorities have existed since World War II, there are recent calls for universities to deepen partnerships with the technology industry and the intelligence community to combat US adversaries in an increasingly competitive global marketplace for AI, big data, and other digital innovations.[31] As universities establish pipelines with government agencies and tech companies to place students postgraduation, and as students are often involved as research assistants and participants, these partnerships directly shape students' understandings of digital tools. Universities are now frequently the site of development for data science practices that disproportionately harm people of color, including predictive policing, facial recognition technology, and risk assessments for the criminal legal system. Smart universities are intimately tied to issues of racial justice, as well as the militarization and corporatization of the university more broadly.

Methods and Goals

The following chapters use historical and textual analysis of higher education administrative discourses, policies, conference proceedings, grant solicitations, news reports, and tech industry marketing materials and product demonstrations to unpack the ways US universities participate in the ethics washing of harmful digital tools that are deployed within and outside of their campuses. Each chapter grapples with a particular set of technologies and how their design, marketing, and reception fit into the larger vision, or sociotechnical imaginary, of smart universities. This focus comes from the understanding that technologies are sites where social, historical, and political forces converge. My hope is that this book will resonate not only with those who are interested in examining technology's impacts on society, but also with those who see the university as a key site of social and political struggle.

This book seeks to actively work against what feminist studies scholars Abigail Boggs and Nick Mitchell have described as the "crisis consensus," the "common sense reflex of liberal investments in the university ... [that] pivots on the invocation of the university as a good in itself, as an institution defined ultimately by the progressive nature at its core."[32] Rather than framing the rise of smart universities as an aberration, I understand these initiatives as a series of new techniques working in service of a long-standing university project: to socialize students in ways that suppress their identification of the university's complicity in structures of domination, including militarism, structural racism, industry ethics washing, and financialization. While much of the book focuses on the impacts of digital technologies on students, it is also the case that many of the forces undermining students' rights in the smart university are the same forces creating conditions of exploitation and precarity for faculty—namely, privatization and austerity.

This study also challenges what American studies scholar Christopher Newfield calls the "austerity consensus," which underpins many smart university initiatives. As Newfield explains, "the core fallacy of the austerity consensus is that there isn't enough money in the economy or public sector to fund public colleges as they once were."[33] Newfield's *The Great Mistake: How We Wrecked Public Universities and How We Can Fix Them* demonstrates that the curtailment of public funding is out of pace with personal income levels, and that the decline of public funding corresponds to the stagnation of graduation rates.[34] Newfield also shows how privatization is driving the increasing costs of public higher education, resulting in tremendous debt for students and their families, the outsourcing of university services to for-profit vendors, administrative growth, and an influx of research that promises financial returns rather than increased public good.[35] Even though public reinvestment in higher education is possible through a progressive tax and the end of private-sector subsidies, the austerity consensus preaches that students should be treated as consumers, and parents, the public, and loan companies as investors. Meanwhile, stagnating wages and student debt compound and intensify existing racial and class inequalities.[36] Discourses concerning the need for and utility of smart universities frequently accept austerity as a given, rather than a set of policy decisions that can be challenged or changed.

Chapter Outline

Chapter 1, "The 'Smart' University," examines what it means for a campus to be "smart." Where does the idea of a smart university come from? What sorts of beliefs, values, and understandings of higher education and technology are embedded in the idea of the smart university? To understand the rise of these initiatives, we must first understand the role of personalization in the digital economy. The promises of personalization for e-commerce are key to administrators' understanding of what makes data analytics attractive for managing both university finances and the student body. Furthermore, the celebratory discourse surrounding smart cities also plays a role in justifications for the use of "smart" technologies in universities. In much of the same ways that smart cities are using data gleaned from parking enforcement, monitoring pedestrian patterns, traffic, crime, and energy usage, proponents of smart universities see a wealth of data that can be used to attract students, improve campus inefficiencies, and address security issues. And, much like smart cities often contribute to the privatization of public space, gentrification, and issues of privacy and discrimination, smart universities can contribute to and intensify these issues as well.

Chapter 2, "Recruitment," explores the parallels between private firms' use of consumer data for targeted advertising to increase the likelihood of a return on capitalist investment and public universities' use of data about prospective students as a way of lowering acquisition costs per student. In both cases, predictive analytics based on demographic and behavioral data, often culled through social media and email tracking, are predicated on significant online surveillance. This approach to student outreach often incorporates data from past examples of successful targeting. In this chapter, I explore how basing predictive models on data from students who were recruited and who successfully graduated can reproduce structurally embedded racial and economic disparities in access. The design of smart recruitment tools often draws from corporate targeted advertising strategies that perpetuate existing forms of racial and economic discrimination in terms of how risks, rewards, and consumer options are distributed online to those whose data profiles fit idealized consumer models.

Chapter 3, "Retention," examines how public universities are turning to big data as a solution to the problem of students dropping out of college. States are increasingly creating performance-based higher education funding systems, which puts pressure on public universities to innovate using predictive analytics. Students are labeled "at risk" based on how their data compares to past students at a given institution. This data is often used to nudge students toward certain majors and courses, as well as to determine when advisor outreach might be needed. In several cases, universities are investing in e-advisors, outsourcing and automating early-stage intervention advising to tech tools. Ultimately, while some of these efforts have indeed increased retention numbers, they have also narrowed students' exposure to different academic disciplines and supported micro-grant approaches to students with financial needs. Rising retention numbers are also used to rationalize the monitoring and management of student behavior in ways that students can experience as invasive, paternalistic, and punitive. Furthermore, this chapter will demonstrate how the data used to train these predictive analytics, much like the data used for recruitment, often reflects historically entrenched inequalities in educational access and support, which then get reproduced and intensified.

Chapter 4, "Wellness," looks at how digital technologies marketed as enhancing students' well-being are part of a longer history rooted in the mental hygiene movement of the early twentieth century, which treated student wellness as a matter of individual responsibility and a failure to adapt to the collegiate environment. This chapter provides a close reading of the WellTrack mobile phone application, a product modeled on cognitive behavioral therapy techniques designed to help users assess and track their own mental health, which offers a real-time dashboard of aggregate student mental health data to administrators. This chapter also examines the integration of Amazon Echo Dots into campus dormitories. It shows how students are being encouraged to turn over their data to privately owned technology firms in exchange for services that will purportedly enhance their well-being. These digital tools construct a vision of a student wellness that precludes an analysis of the structural conditions contributing to poor student mental health, such as austerity and racism, which are social and pervasive, and which would necessarily require an institutional analysis and critique of the university itself.

Chapter 5, "Security," argues that several smart university initiatives are tied to the long-established view of the university as an institution to be secured, rather than a source of students' financial insecurity and vulnerability to state repression. Digital technologies for making universities more secure are symptoms of a pervading discourse of university securitization, which is deployed to justify increased militarized police presence on college campuses. Security discourses contribute to expanding efforts to track students' whereabouts and install security technologies, which double as techniques for producing information about students' utilization of university facilities and for managing university property. Universities have even turned learning management systems (LMSs) into mechanisms for strikebreaking, including for the surveillance and discipline of graduate students withholding grades from the Canvas LMS during the movement for a cost of living adjustment at the University of California, Santa Cruz. Furthermore, this chapter investigates the role of university researchers in advancing digital technologies like artificial intelligence in partnership with government agencies and private industry for the specified purposes of national security. It analyzes the outcomes of these partnerships on structurally disadvantaged communities both within and outside of the academy.

In Chapter 6, "(Beyond) Digital Literacy," I examine smart university proponents' claims that today's students are digital natives. Digital natives theoretically grow up with a greater familiarity and understanding of technology relative to those born prior to its widespread adoption. This position naturalizes how smart universities socialize students to engage with technologies in ways that promote their identification with the expectations, rhythms, and demands of the digital economy. Instead, students need digital literacy training that is specifically geared toward the university, encouraging critical thinking about how technologies mediate their educational experiences in ways that entrench austerity and racial injustice in US higher education. One of the ways this literacy can be cultivated is through greater knowledge of digital tools' social and ethical impacts within universities. While calls for digital literacy have in many cases been for the purposes of retraining students to meet new workforce demands, the digital literacy called for in this chapter is fundamentally critical, provoking students not only to question their personal use of digital technologies, but also to see themselves as situated and technologically mediated actors within the university. This chapter draws from how instructors are currently engaging with AI, participatory design, and satire to imagine critical digital literacy for the smart university.

Yet greater student and faculty digital literacy about smart university initiatives alone is unlikely to fundamentally transform higher education, given the university's long-standing complicity in structural inequality. The conclusion, "Against the Smart University," argues that we need a coalition-based movement against the encroachment of the private tech industry on the university and the conditions that facilitate it. Students and faculty together could refuse to participate in the production and use of digital tools that harm marginalized people within and beyond the academy, and mobilize against the institutional politics of austerity. These efforts are already taking shape: students are organizing against the remote proctoring software, Proctorio, given its invasive digital surveillance, inaccuracies, and punitive consequences for disabled students and students of color, and they are protesting faculty participation in predictive policing technology research. Mathematics faculty from several institutions are refusing to collaborate with police departments in the wake of nationwide demonstrations against anti-Black racism and police violence. There are widespread movements for the full cancellation of federal and private student loans for college. Coalitions formed across all levels of the academy and in solidarity with grassroots community organizers can hold universities to account for their past and present-day participation in the structural violence they claim to redress. This requires abandoning the notion that universities are victims of neoliberalism rather than long-standing mechanisms for perpetuating injustice, including through the development and deployment of digital technology for making universities "smarter."

Chapter One

The "Smart" University

Imagine, for a moment, a first-year student named Yasmine studying in her dorm room with her roommate, Jessica. The university's roommate-matching algorithm recommended Jessica to Yasmine based on their compatibility score. This score was derived from a comparison of their self-reported habits, interests, and personality test outcomes. Furthermore, the university's recruitment outreach and financial aid offer to Yasmine hinged on a range of collected data points about her, including her demographic information, social media engagement, and web browsing behavior. These data points were fed into the university's predictive analytics, which determined that Yasmine's data profile was similar enough to past successful recruits to be a cost-effective target. Nervous for her second full week of classes, Yasmine logs moderate stress levels on the free-to-use mental health self-tracking app that her university subscribes to, which introduces Yasmine's health data to third-party services. As Yasmine heads to the dining court and then the library, the swipes of her student ID card track her location. She then logs into her university's learning management system to check her grades, and the platform captures her login time, duration, and location too. When she finally arrives to her classroom, a Bluetooth beacon logs Yasmine as present, having detected her cell phone. These data points will join a sea of insights that Yasmine's smart university will use to predict, anticipate, label, and nudge her behavior toward graduation.

But what does it really mean for a university or campus to be "smart"? International firms like Deloitte market these initiatives as frictionless, seamless, and a means of continuous campus modernization. Smart campuses are, according to proponents, a way to leverage next-generation technologies already transforming commercial industry and city governance for the better.[1] In some cases, universities have partnered with technology firms in order to serve as a proving ground for new technologies and offer technology workforce training. One such example is Northern Arizona University's (NAU) partnership with the telecommunications company CenturyLink. In February 2019, NAU and CenturyLink created a lab for students and faculty to investigate how towns, cities, and campuses use cutting-edge smart technologies. Regarding this initiative, NAU's vice president for information technology and chief information officer stated that the university has "an obligation as a higher education institution to foster those advancements that improve the human condition."[2] In this sense, the smart university becomes a moral imperative.

Furthermore, smart university initiatives in the United States dovetail with efforts underway in several states to link K–12, higher education, and workforce data to track student performance "from cradle to career," entrenching the notion that education is fundamentally about workforce preparation.[3] While the focus of this book concerns smart university initiatives in the United States, several universities in Latin America, the United Kingdom, South Korea, Spain, and Australia have been embracing the idea using a range of approaches.[4] There is also rising interest on behalf of OECD

(Organisation for Economic Co-operation and Development) countries to measure and target the psycho-emotional aspects of education in order to nudge students to develop the personality traits and socio-emotional qualities believed to lead to "grit," "resiliency," and "robot-proof" skills in the face of volatile labor markets. These efforts are largely based on an assumption, rooted in behavioral economics, that students will not make rational self-interested decisions for sufficiently meeting academic demands, and instead behave according to noncognitive emotional and psychological processes.[5]

According to the marketing rhetoric for smart universities, while students might not always behave rationally, they are largely tech-savvy, always-connected, digital natives. They are consumers to be courted, future workers to made employable for increasingly smart workplaces, sources of user-generated content for marketing and outreach, and resources to be mined for making campuses even smarter.[6] In some cases, increasing enrollments from students with "nontraditional" backgrounds is used to create a sense of urgency around smart university initiatives for student success, where monitoring and nudging these students to persist is marketed as a solution to the problem that most universities are fundamentally not built to support them.[7] The term "nontraditional students" is typically used to refer to students who are financially independent, enrolling one or more years after graduating high school, attending part time, or working a full-time job while in school. It may also refer to students who have children, first-generation college students, or those with a high school equivalency diploma.

While some proponents distinguish between smart universities and smart campuses, the former focusing on infrastructure and academic services and the latter on economic and financial perspectives, many proponents use the terms "smart campus," "digital campus," "intelligent campus," and "smart university" interchangeably.[8] I use the term "smart university" to include industry providers of digital technologies for higher education governance that might market their services with a range of technocentric terms but nonetheless make related promises to optimize university resources and enhance prospective and current student experiences using advanced digital tools and data-informed insights that purportedly combat uncertainty and enhance institutional competitiveness.

The celebratory rhetoric surrounding these initiatives primarily draws from two preexisting industry-related phenomena: personalization and smart cities. Both have been sites of rigorous critique for their intensification of existing inequalities in society. This chapter will demonstrate how smart university initiatives draw from the celebratory rhetoric and technocratic underpinnings of personalization and smart city initiatives, promising to provide customized care to students and maximum efficiency for managing university resources. First I unpack precisely what personalization is, how it operates in the digital economy, and how universities take up this strategy. Then I examine how smart university initiatives leverage the discourses and strategies of smart city initiatives.

Brief History of Personalization

According to some smart university advocates, appropriating the data mining and predictive analytics strategies of industry is key to identifying students' needs and enhancing "institutional productivity," "student success," and "operational decisions" while creating personalized, targeted, real-time interventions.[9] In the context of the digital economy, personalization is the customization of advertisements and services based on the digital surveillance of consumer behavior. Generally speaking, surveillance can be defined as "the operations and experiences of gathering and analyzing personal data for influence, entitlement, or management."[10] With personalization, data analytics are used to structure how goods, services, and content are distributed to targeted audiences. It can be considered a form of "dataveillance": using personal data to not only monitor consumer behavior but also predict and prescribe.[11]

Personalization is indebted to the historical development of statistics for managing populations. According to sociologist Armand Mattelart in *The Information Society*, advancements in statistics and probability theory were essential for the rise of the welfare state, and ultimately, the information society.[12] Technologies for governing and regulating society, including assessing risk through statistical probabilities and marking certain populations as criminal or deviant, migrated from technologies for governing disruptive populations under the welfare state in the twentieth century to modes of governing the market under neoliberal capitalism.

The development of personalization was contingent upon the migration of data analysis from the state to commerce, and the emergence of digital marketing in the 1990s. Oscar Gandy's renowned sociological study of consumers' relationship to information collection led him to conclude that their decreasing willingness to participate in surveys and rising consumer rights activism in the 1970s–1990s meant new, targeted marketing methods had to be devised. These methods included collecting, packaging, and selling consumer information and targeting consumers through phone calls, email, and eventually, online browsing.[13]

The rise of digital marketing went hand in hand with the use of advanced analytics to measure consumer behavior. Data analysis as a discipline arguably began in the 1960s with the fusion of statistics and computer science and the establishment of the International Association for Statistical Computing in 1977. By 1989, databases and data mining became an emerging field of social and technological inquiry and put into motion the discourse that would eventually solidify around "big data." Whereas data aggregation was initially dependent on the manual accumulation of sales data from cash registers, inventories, and consumer polling, the development of the bar code, and later, sophisticated computing technologies, enabled major retail companies to reduce labor costs and leverage knowledge of the market against suppliers, competitors, and consumers to control costs. This shift allowed retailers to demand greater flexibility from manufacturers and to control labor flows.[14]

By the 1990s, the expression "data mining" became popularized in mainstream culture, and by 2005, companies would begin competing using extensive analytics and algorithms to mine data and produce valuable information for managing warehouses, transportation infrastructure, and industrial rhythms. In response to the dot-com bubble bursting in 2000, targeted advertising served as the foundation for a new capital accumulation strategy. The federal government supported these tracking efforts by permitting the web to be restructured around the capture of user data, including through cookies, location data, and search histories, for digital services and advertisements.[15]

User behavior is made legible to personalization technologies through soliciting and monitoring user attention. The increasing capacity of the web to be interactive is what is considered to differentiate web 1.0 from web 2.0, where web 2.0 designates the period when users began directly interacting, connecting, and participating on the web with one another and with businesses. None of this would have been possible without the developments of relational databases in the 1960s and '70s, which allowed data to be stored in ways that made searches efficient. Today, with cloud computing, data can be preserved and aggregated using a variety of methods. These developments allow for the accumulation of large swaths of data that would otherwise be considered valueless, unstructured, and unquantifiable if it were not for the software that allows it to be turned into information.

Personalization algorithms are one of the means by which data is turned into information: a step-by-step computation used for processing data. As communications scholar Stephanie Ricker Schulte explains, "People often think of algorithms as autonomous and impersonal, but they are in actuality the material process of personalization, the codification of assumptions about persons, the values held by persons, into computing systems of quantitative processing."[16] These assumptions include but are not limited to the idea that the collected data represents people accurately, and that it can serve as a sound basis for distributing options and choices to different individuals. However, market forces are often driving who implements personalization and how.

From Netflix and Amazon recommendations to banking, insurance, and social networking, personalization is used to increase the likelihood of a return on financial investment using predictive analytics. In the context of e-commerce, proponents argue that personalization provides individually tailored and interactive goods, content, and services, thereby overcoming the limits of mass culture's homogeneity and standardization. Consumers are theoretically empowered through their capacities to interact with platforms and to have their desires anticipated.[17] Media studies scholar Mark Andrejevic describes this discourse as the now-it-can-be-told promotional strategy of the digital economy, which admits to the repressive limitations of mass media in order to tout the promise of interactivity: "It turns out that critical theorists were right about industrial capitalism all along: it is oppressive, top-down, and alienating after all. We can finally admit this because we now have the technology to leave it all behind."[18] While Frankfurt School theorists like Theodor Adorno and Max Horkheimer argued that under capitalism, mass culture produced conformity under the guise of freedom of choice, personalization purportedly empowers consumers with more agency than ever before because their preferences and behaviors impact what goods and services are marketed to them as most relevant.[19]

Critics have pointed out a range of harms that can result from personalization online, from diminished privacy rights and the consolidation of corporate power over user data, to social and economic discrimination.[20] It's important to note that personalization, like other modes of surveillance, often results in discriminatory treatment, helping to concretize race and other markers of social difference as categories that can be implicitly or explicitly employed to target users. In this sense, personalization can be described as a form of racializing surveillance. Surveillance studies scholar Simone Browne defines racializing surveillance as structuring "social relations and institutions in ways that privilege whiteness."[21] As just one example, in 2022, the US Department of Justice sued Meta Platforms, Inc. (formerly Facebook, Inc.), for violating the 1968 Fair Housing Act, given the ways its "Special Ad Audiences" advertising algorithm enabled discrimination against users based on race, gender, and other characteristics for targeted advertisements. Furthermore, both Staples' and Home Depot's personalization tools have charged customers in poorer zip codes higher prices than those in wealthier zip codes.[22] Yet, according to proponents like Hal Varian, Google's chief economist (who developed the algorithm for their advertising system, AdWords), differential pricing is egalitarian:

> Forcing a producer to sell to everyone at the same price may *sound* like a good idea. But it can easily end up encouraging the producer to sell only to the high end of the market. Differential pricing gives the producer an incentive to supply the product to everyone who is willing to pay the incremental cost of production... Forcing a policy of flat pricing in an industry where it is inappropriate due to the nature of technology may well have perverse consequences.[23]

Varian's argument is that differential pricing allows for the majority of consumers to enjoy the same goods by correlating the price of a product to the consumer's means. This model of differential pricing is dependent on the collection of data about consumers to determine the highest price each consumer would be willing to pay.

Varian's egalitarianism neglects to account for how personalization can contribute to "digital redlining," a term privacy scholar Chris Gilliard coined to describe "the creation and maintenance of technological policies, practices, pedagogy, and investment decisions that enforce class boundaries and discriminate against specific groups."[24] In this sense, differential pricing can be situated within a wider set of discriminatory practices that use information about people to manage the options and choices they are presented with, particularly to incentivize those most likely to provide a return on capitalist investment and to manage the kinds of services and options, or lack thereof, offered to those deemed too risky.

Digital redlining is indebted to the practice of redlining, where the US federal government and lenders refused to invest in neighborhoods based on their racial or ethnic makeup, first formally institutionalized in the National Housing Act of

1934.[25] The Fair Housing Act of 1968 was supposed to redress this practice in the context of mortgage lending and real estate. Not only did this act fail to integrate the housing market, but civil rights era protections more broadly have proven inadequate for addressing algorithmic discrimination in the present. With big data, service providers can actively limit choices and price goods for different consumers, including in contexts that fall outside the purview of anti-discrimination law. Those seen as worthy of investment are presented with better options, incentives, and prices, while those who are marginalized, perceived as risky, or whose data doubles do not appear as viable opportunities for creating profit are presented with different choices that can be prejudicial or exploitative. Despite the common refrain that algorithmic models are neutral and objective, in practice, they often reinforce and amplify existing social hierarchies. As just one example, computer scientist Latanya Sweeney's study from 2013 demonstrated that web searches involving Black-identifying names were more likely to display ads with the word "arrest" in them than searches with white-identifying names.[26] Low-income students and students of color are frequently the target for predatory loan advertising on platforms like Facebook. These predatory practices for tracking and targeting marginalized people were also a contributing factor to the subprime-mortgage bubble and ensuing financial crisis in 2008.[27] Yet to this day, whether people are pursuing jobs, higher education, or searching for goods and information online, personalization can cause immense harm.

Given that personalization can amplify inequities of access to rights and resources, it cannot be understood as a tool for simply presenting people with the best options on the market or the most relevant information for their preferences and desires. Rather, it is part of a larger sociotechnical system that aggregates data for the purposes of assigning risk and opportunity. It reinforces long-standing discriminatory practices prevalent in industries like banking and insurance to maximize profit.[28] Additionally, personalization technologies have implicit assumptions about how to formulate categories that can intensify marginalization. For example, in 2009, more than fifty-seven thousand gay-friendly books disappeared from Amazon's sales lists because they had been categorized, however inadvertently, as "adult." As communications scholar Tarleton Gillespie explains, "naturally, complex information systems are prone to error. But this particular error also revealed that Amazon's algorithm calculating 'sales rank' is instructed to ignore books designated as adult. Even when mistakes are not made, whatever criteria Amazon uses to determine adult-ness are being applied and reified—apparent only in the unexplained absence of some books and the presence of others."[29] Thus, the use of personalization to manage the marketing and distribution of consumer choices is inextricably linked to normative assumptions about certain kinds of content: how it should be categorized, organized, censored, or displayed.

This analysis of personalization provides a much more complicated picture than Bill Gates's "frictionless capitalism," a free market utopia of perfect information where everyone in the market is able to make informed decisions and where society's resources are distributed evenly.[30] Gates has also been a longtime influential proponent of student tracking, arguing in favor of student performance monitoring from childhood through high school and college and into the workplace at the 2009 National Conference of State Legislatures.[31] Despite concerns regarding privacy and discrimination, personalization is foundational to many smart university initiatives, from determining which students are most in need of advising, to managing enrollment, to assessing which students are most likely to default on student loans.

From E-commerce to Higher Ed

In the context of higher education, personalizing student experiences includes leveraging their data for targeted recruitment and retention efforts. Data about prospective and current students is used to nudge their behaviors and choices to align with university metrics of success. While economist Richard Thaler and legal scholar Cass Sunstein conceptualized nudges as incentives designed to preserve an individual's freedom of choice, nudges here are best understood as automated techniques that order and pattern what seems possible and impossible for students.[32] Many present-day university approaches to personalization, such as Austin Peay State University's course recommendation system Degree Compass, are explicitly marketed as being modeled after the personalization techniques of companies like Netflix and Amazon. For instance, whereas Amazon uses a collaborative filtering algorithm to compare users' purchasing habits to those of other customers to make product recommendations, Degree Compass compares past students' grades to current students' transcripts to make tailored course recommendations.[33]

Just as industry personalization claims to center individual consumer needs, higher education personalization claims to center individual student needs. Yet, in both cases, predictions about what a given consumer or student needs are predicated not on individualized treatment but, rather, on finding patterns in mass aggregate data to suit predetermined goals. Similarly, while proponents of industry personalization appropriate critiques of mass culture to market predictive analytics, proponents of personalization for higher education appropriate critiques of schools as outmoded factories of standardized mass education to frame the need for personalization technology.[34] And, much like major media technology conglomerates use nudges to get their users to engage with their platforms in ways that increase their market power, universities use nudges to consolidate control in a volatile higher education marketplace.

The idea of using technology to "personalize" learning emerged well before the rise of Silicon Valley in the United States. Education technology expert Audrey Watters has traced present-day discussions about personalized learning and computing technology in education to the twentieth century's teaching machines movement.[35] Early twentieth-century proponents of teaching machines like Benjamin D. Wood, an American educator, researcher, and professor at Columbia University, argued that these tools could eliminate drudgery for teachers and offer students more individualized, albeit automated, instruction than could standardized mass education. As Watters explains, this individualization of education was predicated on the idea that it was necessary to *know* students through testing—content examinations, psychological analysis, personality assessments, intelligence and aptitude tests—at all stages of their learning, from kindergarten to university.

Personalization, Watters teaches us, has "long meant data collection and analysis."[36] In the mid-twentieth century, educators' resistance to teaching machines, which included concerns about the automation of the profession, its alienation of students from teachers, and student dehumanization, were often met with critiques from industry proponents. These proponents argued that teachers were inefficient, averse to technological progress, and resistant to change.[37] The idea that education needed to be optimized and mechanized intensified after the Soviet Union's successful launch of the first artificial earth satellite, *Sputnik*, in 1957, leading to approaches among major funders like the Ford Foundation that privileged the expertise of engineers and technologists rather than teachers when it came to reforming education.[38] This industry-oriented, technocentric consolidation of power over the future of education continues today.

An additional parallel between present-day personalization for e-commerce and for higher education is the use of historical data to forecast, and in turn attempt to manage, the future. An example of this in the context of industry is Amazon's patent for anticipatory package shipping, a system where products are shipped to warehouses located near customers who are predicted to buy the goods. The system would base this prediction on an algorithm that uses data from customers' previous searches, purchases, preferences, and the length of time an individual's cursor hovers over a certain image.[39] For smart university proponents like Zhao Yang Dong, Yuchen Zhang, Christine Yip, Sharon Swift, and Kim Beswick, "The enormous educational data generated from the information system offers opportunities to learn from the historical observations and forecast future conditions ... The forecast ability of a smart campus can provide a more manageable learning environment and promote forward-looking ability."[40] In both the case of e-commerce and higher education, data is being used for preemptive decision making that impacts distribution of resources and opportunities in these varied contexts.

Chapters 2 and 3 will demonstrate in greater detail how the use of predictive analytics for student recruitment and retention is often structured around the priorities of increasing tuition revenue and institutional rankings under conditions of austerity, reinforcing long-standing inequalities in which students are targeted for enrollment and intervention. Rather than solutions that substantially increase student resources or involve policy shifts that make universities accessible for all, smart university initiatives encourage administrators and higher education professionals to leverage digital tools inspired by and often directly procured from industry to automate students' trajectories using digital forms of surveillance and control. Underpinning the rhetoric of personalization is a fundamental depersonalization of the university, a hostility to the complex, interpersonal, and dynamic work of education.

From Smart Cities to Smart Universities

The ideological roots of smart universities can also be found in visions and applications of smart city initiatives. For instance, in 2014, the University of Lille presented a case study to the World Bank to illustrate how the smart campus could help realize the concept of the smart city.[41] Similarly, a white paper from software development company Collabco argues that US universities are ideal places for researching and developing smart city technologies, providing the necessary research infrastructure and opportunities for prototyping and testing on campus while "attracting some of the brightest students to their institutions."[42] Although there is no universally agreed-upon definition of a smart city, smart cities generally use knowledge acquired through the diffusion of information and communications technology (ICT) for urban governance, often for the espoused purposes of increased sustainability, resource efficiency, innovative commerce, and a higher quality of life.[43] Smart cities promise to use more objective and neutral technological solutions to solve urban problems, using data to anticipate uncertainty for city governance.[44] In practice, however, many smart city initiatives contribute to gentrification, the privatization of public space, pervasive data collection for policing, and corporate control over civic processes.

As just one example, in geographer Christopher Gaffney and communication researcher Cerianne Robertson's analysis of Rio de Janeiro's smart city initiative, which emerged in tandem with Olympic planning agendas in 2016, they point out that smart city technologies are not being used to address problems of systemic poor governance or social inequality.[45] Similarly, Alphabet Inc.'s smart city initiative in Toronto promised to use technology to create a sustainable, data-managed urban neighborhood and yet failed to incorporate substantial planning for affordable housing or transparent and accountable data-collection plans. Privacy concerns, particularly concerning third-party access to people's identifiable information, led former Ontario privacy commissioner Ann Cavoukian to resign from the project. The #BlockSidewalk campaign was instrumental in organizing around not only these privacy issues but also a lack of community consultation and the ceding of public land to a private corporation. Ultimately, the project was canceled.[46]

Many proponents of smart cities obfuscate these concerns when they focus on their purported benefits, including "convenience," "efficiency," and "seamlessness." These terms feature prominently in IBM, Siemens, and Microsoft smart city literature. According to urbanist Adam Greenfield, "seamless" in this context means that "the user perceives no interruption in the flow of a technically-mediated experience ... [T]he language of seamlessness implies that the hassles of everyday life have been mitigated by the intervention of powerful technologies, from whose complexity, in turn, the user has been carefully and deliberately shielded."[47] Seamlessness in the smart city comes at the expense of people having insight and control over how these technologies are mediating their lives.

Much like the celebratory rhetoric surrounding seamless smart city technology, proponents of smart universities argue that they provide "frictionless, touchless, and intuitive experiences driven by a digitally connected community."[48] Some proponents, such as the private consulting firm Deloitte, explicitly link smart campuses to smart cities in terms of maximizing operational efficiencies:

> Being a smart campus enables institutions to streamline processes, reduce costs, and become operationally more effective through insight, automation, and the replacement of antiquated procedures with innovative practices. Campuses are able to use insights gained from a data strategy to proactively address issues, enabling efficiency through a thoughtful redesign. Similar to a city, maintaining buildings, facilities, landscapes, and power grids has become vital to campus's operational efficiency.[49]

One campus that *Guardian* writer Zofia Niemtus frames as a "mini" smart city approach to sustainability is the University of Texas at Austin, given its use of an independent grid powered by natural gas turbines to reduce energy costs and carbon dioxide emissions.[50] Yet, UT Austin is one of the largest university landowners in the United States, with much of that land leased to the oil and gas industry. Claims about university sustainability must be assessed not only in terms of their use of data and greener infrastructure to reduce their own emissions but also in terms of their wider relationships to fossil fuels, including through land leasing, investment-related assets, and donations from Big Oil companies.[51]

Parallels between smart universities and smart cities go beyond matters of resource and facility management. Deloitte also imagines the use of advanced digital tools like facial recognition technology, reporting and alert systems, gunshot sensors, and location intelligence to keep campus communities safe similarly to how cities are increasingly being policed. Stephanie Weagle, the chief marketing officer at BriefCam, a leading industry provider of video analytics for rapid video review and search, real-time alerting, and quantitative video insights makes this comparison explicitly:

> Today, educational institutions operate much like a city, with centralized law enforcement ensuring the safety of students and visitors, responding and investigating when crime occurs and striving for improved quality of life for residents ... [S]ecurity surveillance is used across campuses, but with the addition of artificial intelligence and data-driven video analytics and real-time alerting capabilities, security can enhance situational awareness, detect unusual and excessive dwelling and accelerate responses to emergencies, threats and suspicious behavior.[52]

Several universities have already begun integrating AI and other digital tools into their campus security measures, discussed in greater detail in chapter 5. Yet smart city policing measures have been the subject of significant scrutiny, including for a lack of transparency and democratic oversight, police militarization, sweeping expansions of state surveillance over quotidian activities, and the intensification of policing in communities of color.[53]

Automatic license plate readers, StingRay devices that mimic cell phone towers to log nearby phones, and algorithms like PredPol—a policing tool for predicting crime "hot spots"—are all examples of smart policing technology that dramatically expand state surveillance powers. This technology is well known to perpetuate the racial injustice baked into the existing criminal legal system, one reason being that the data used to train predictive policing algorithms emerges from a context where poor people and people of color are disproportionately targeted and criminalized.[54] Many for-profit companies develop smart policing tools with funding from the US government that is then sold to city police departments, oftentimes with the participation of university researchers. Despite documented harms of racial profiling and discrimination, smart university initiatives continue to draw upon these digital policing strategies to secure their campuses. According to Weagle, these aspects of smart campuses serve the dual purpose of helping to improve traffic and visitor movement optimization, and in turn, "efficient energy consumption across campus,"[55] greenwashing these discriminatory policing methods. Similarly to how tech-driven smart cities often expand surveillance and privatization with little democratic oversight, smart university initiatives can intensify the tracking of campus life and privatize higher education governance without accountability.

An additional parallel between smart universities and smart cities is the constitutive role that narratives of crisis play in the proliferation of smart technology. Vendors use narratives of financial, demographic, and ecological crisis to sell smart technology solutions to local governments.[56] In Western cities, smart technologies for urban governance emerged in the wake of long-term recession and austerity measures, with companies promising to help city governments improve inefficiencies, address outdated and unsustainable infrastructures, and cope with limited financial resources. For smart universities, the narratives of crisis deployed in service of marketing smart technology for higher education include austerity and the shifting demographics of prospective student bodies. Siemens, for instance, markets its solutions as a means of addressing higher education's "increasingly diverse student demographics" in a context where "public funding is in a stalemate or has even been reduced."[57] The COVID-19 pandemic has only intensified anxieties about projected US higher education enrollments, down by 1.3 million students since spring 2020.[58] In both the case of smart cities and smart universities, digital technology is being used to address crises constructed in marketing literature to increase the power and control of industry forces.

Finally, in both the case of smart universities and smart cities, proponents sometimes gesture toward the need for community input to make sure these initiatives are responsive to the people they will directly impact. However, critics of smart cities have argued that community participation is often limited and perfunctory, primarily serving to manufacture consent for predetermined initiatives. For instance, in the case of Alphabet's smart city project in Toronto, the company's urban planning and infrastructure subsidiary, Sidewalk Labs, partnered with the Inclusive Design Research Centre at the Ontario College of Art and Design University on eight codesign sessions that were held with community groups across Toronto. Sidewalk Labs also held three daylong sessions with people who identified as part of the accessibility community. Yet the project had little baked-in democratic oversight over its proposed corporate takeover of public land, resources, and data. Furthermore, it aimed to exceed its original 12-acre allotment to 250 acres, according to leaked documents.[59] Smart city infrastructure providers seldom go into detail about the degree to which their algorithms or designers will face processes of democratic accountability.

Similarly, some proponents of smart universities argue that these initiatives require participation and commitment from multiple stakeholders, including students, faculty, staff, and parents, to be successful.[60] Rarely are people outside the university envisioned as key stakeholders, despite the fact that universities dramatically impact the living and working conditions of those beyond their walls, from housing costs and employment opportunities to policing. Furthermore, many proponents do not indicate what, exactly, gathering stakeholder input means in practice. In some cases, stakeholder input consists of data collected from anonymous surveys intended to mitigate concerns, rather than a substantial democratic process for building consensus about what (if any) problems smart university tools should be used to address and how.[61] This lack of democratic design and governance is consistent with science and technology studies scholar Jathan Sadowski's analysis of smart technology more broadly, which he argues "advances the interests of corporate technocratic power, and over other values like human autonomy, social goods, and democratic rights."[62] Smart university initiatives vastly expand the powers of industry over higher education governance, modeling and in many cases directly commissioning its strategies for surveillance, exploitation, and control.

At their core, smart technologies, be they for department stores, cities, or campuses, rely on vast amounts of data. This raises pressing concerns regarding people's abilities to control who knows what about their activities, locations, preferences, and relationships. There is no shortage of examples of data from smart technologies being used to widen the reach of corporate and state power. As just one example, Amazon's smart doorbell, Ring, has provided doorbell footage to police without owner consent or a warrant, contributing to a vast police surveillance network over residencies.[63] However, these sorts of digital tools raise additional issues beyond privacy, which is often narrowly construed as having control over one's own information. It is always the case that technology involves embedded values, goals, and assumptions about the world, including how problems should be solved, how decisions should be made, and how advantages and disadvantages should be distributed in society. In this sense, and as many science and technology scholars have argued, technology is not neutral but fundamentally political.[64] Similarly, as media historian Lisa Gitelman and literary scholar Virginia Jackson have argued, data is not neutral or "raw"; rather, it emerges from an existing social, political, and historical realty under conditions of observation that have been created according to particular needs and uses.[65] Smart university initiatives produce and collect data for the purpose of guiding student choices and logistically managing campus life. Complex challenges facing higher education are abstracted and simplified so that they can be managed using technological solutions.

As just one example, in sociocultural anthropologist Madisson Whitman's ethnographic study of predictive analytics for tracking student success, the university's data personnel chose to remove demographic markers from their predictive models. Instead, they focused on data regarding individual behaviors, which students can theoretically change. These behaviors included engagement with learning management systems and class attendance. However, as Whitman explains, focusing on individual behaviors "enables the institution to transfer the burden of success away from itself and keep the tacitly held knowledge of inequality out of the university's visions for predictive modeling."[66] Leaving out demographic markers allows institutions to claim they are using unbiased data, despite the fact that students' life chances and experiences in higher education are dramatically shaped by how their race, gender, ethnicity, (dis)ability, and socioeconomic status intersect with matters of access, opportunity, and support.

This framing of data as unbiased is also prevalent in smart city discourse. As Greenfield points out, "When discussing their own smart-city venture, senior IBM executives argue, in so many words, that 'the data is the data': transcendent, limpid, and uncompromised by human frailty."[67] In both the case of the smart city and of the smart university, complex circumstances are often measured using more easily determined proxies so that digital tools are able to observe and address a given phenomenon. Despite claims that the data used for smart university initiatives is unbiased, these tools are designed to mobilize information to suit institutional ends in ways that do not fundamentally trouble the status quo.

What makes these digital technologies so seductive to higher education administrators, in addition to promises of cost cutting, enhanced student experiences, and improved rankings, is the notion that the integration of digital technology on their campuses will position universities to keep pace with innovation. Some economists like Andrew G. Haldane refer to the present as the Fourth Industrial Revolution, characterized by the confluence of AI, big data, automation, nanotechnology, and the Internet of Things. Like moments of industrial revolution before it, the thinking goes, where technological transformation also created periods of disruption and widening inequality, Haldane sees institutions as having the power to train workers to meet new demands and cushion the socioeconomic impact of new technologies. For Haldane, the role of universities in the Fourth Industrial Revolution, or what he calls "multiversities," is to facilitate combinations of skills that cannot be automated. Haldane also sees multiversities as sites for diffusing technologies through partnerships with business networks.[68]

Several US higher education administrators who argue for the increased adoption of digital tools for monitoring and managing student success simultaneously argue for deeper partnerships between universities and the private tech industry. They see these partnerships as a means of training students to meet new workplace demands while staving off automation, unemployment, and decreased wages. For instance, Joseph Aoun, president of Northeastern University, argues that higher education needs to be transformed in order to offer skills that are robot-proof, meaning skills that will not be replaced by technological advancements in AI.[69] His book sees potential in both edtech and the gamification of student learning, as well as cooperative education and research partnerships between businesses and universities. Michael Crow, president of Arizona State University (ASU), and historian William B. Dabars similarly argue in *Designing the New American University* that universities need to transform in order to address economic constraints on public funding and the inadequacies of current university models for providing socially relevant education for the global knowledge economy. The text argues for corporate partnerships as being "essential if universities are to leverage their knowledge production and catalyze innovation that benefits the public good."[70] Crow's approach to overhauling ASU, developed in part as a response to its defunding, has been met with criticism concerning the corporatization of the university, the stifling of academic freedom, and an overreliance on technological solutions in the classroom.[71]

The discourse of the Fourth Industrial Revolution is also fundamentally technologically deterministic in framing these innovations as inevitable, deferring to market-driven processes. It has largely been constructed and propagated by consultancies, think tanks, and economic modelers, including the World Economic Forum's Klaus Schwab, McKinsey under the leadership of James Manyika, and Deloitte.[72] It has been influential in the development of smart universities both within and outside of the United States, including the EU, South Korea, and Saudi Arabia. Information technology experts from across these regions have argued in their scholarship that smart campuses are needed to best prepare students to live and work under the Fourth Industrial Revolution.[73]

Schwab was the first to introduce the idea of a Fourth Industrial Revolution at the World Economic Forum in 2016, drawing from a 2011 strategic initiative of the German government to increase digitization in manufacturing.[74] Workers (and students understood as workers in training) are positioned within this discourse as responsible for shoring up future employment in a context where the development and consolidation of networked technologies is unstoppable, and for some proponents, perhaps even the key to ending inequality. Little attention is given to evaluating the socioeconomic environment that gives shape to these innovations. For instance, economic historian and social theorist Aaron Benanav has shown that productivity and economic growth have been decelerating globally for decades.[75] Rather than a set of recent, unprecedented shifts in technology and labor, the present is being shaped by longstanding historical conditions of global exploitation and automation under capitalism, including the structural relationship among globalization, neoliberalism, and the proliferation of networked information communications technology beginning in the 1970s.[76]

It is these more long-standing transformations that have led to ongoing job losses, the decline of the middle class, and precarious working conditions in the global economy. These are social conditions that the vast majority of students will not escape through individual educational attainment. Instead of students being empowered to imagine alternatives to the prevailing social and economic order, Fourth Industrial Revolution discourse emphasizes students' need to ensure they can meet the demands of data-intensive industries or otherwise face un- or underemployment. This chapter has argued that proponents of smart universities leverage the celebratory rhetoric and strategies of personalization and smart cities to position smart universities as solutions to higher education challenges, including maintaining the public university's relevancy in a period of rapid technological change and diminished public funding. The logic of the smart university prioritizes risk mitigation for the university, while preparing students to live and work in a digitally mediated world where little about their future is guaranteed. In the next chapter, I examine in greater detail how universities in the United States deploy "smart" data analytics strategies, like personalization, to target and recruit these students.

Chapter Two

Recruitment

In March 2019, the Department of Justice charged fifty people with fraud for their participation in a college admissions scheme that would help wealthy people's children gain entry to elite universities by falsifying their qualifications. Through the Key Worldwide Foundation, a sham charity that claimed to serve underprivileged students, the money of wealthy parents was laundered to bribe test administrators of the American College Testing (ACT) and Scholastic Aptitude Test (SAT), as well as athletics coaches. Ultimately, parents were able to get their children admitted to colleges including Yale, Stanford, Georgetown, the University of California, Los Angeles, the University of Southern California, and the University of Texas at Austin. They used falsified test scores, doctored photographs of their children playing competitive sports, exploited disability accommodations for standardized exams, and lied about students' racial and ethnic backgrounds. There was no shortage of public backlash against the high-profile elites involved, and headlines frequently emphasized the illegality and unfairness underpinning this scheme.[1] Many coaches have since been fired, some parents received plea deals or were convicted of crimes resulting in fines and imprisonment, and several students have had their admissions revoked.

While the use of fraud and deception to manipulate the higher education admissions process is objectionable, public fixation on the unfairness of these parents' actions suggests that the admissions process is, itself, otherwise fair—the problem is when people abuse their wealth or influence to game an otherwise meritocratic system. However, what the history of higher education admissions reveals is that many universities actively participate in the racial and economic stratification of resources and opportunities in society in ways that are fundamentally unfair, inequitable, and contingent on institutional surveillance. This chapter demonstrates how smart university initiatives for recruitment risk continuing these longstanding patterns of exclusion and enforce a politics of top-down austerity management at public universities. First, this chapter provides a brief overview of admissions in US higher education and the ways that race, gender, and class have shaped admissions practices historically. The latter half of the chapter examines how emerging forms of digital tracking are shaping the goals, processes, and disparate outcomes of recruitment and admissions in contemporary public universities.

A Brief History of Admissions in US Higher Education

The Morill Acts of 1862 and 1890 provided federal funds and lands for the development of US public colleges, primarily in the Midwest and West. In contrast to early private colleges, which offered religious and legal training, land-grant colleges were seen as key for expanding US economic interests in agriculture, science, and industrial technology. The development of these public institutions hinged on the violent expropriation of Indigenous lands and enforced racial segregation by requiring separate colleges for Black students.[2] People who attended college during this period were typically white men who came from wealthy and established families and pursued specialized professional training. Although the demand for college prior to the 1940s was not high, admissions officers actively discriminated against applicants of color as well as Jewish applicants, and white women were generally able to attend only women's colleges.

In 1944, the Servicemen's Readjustment Act, also known as the GI Bill, dramatically increased enrollments in higher education, leading to the establishment of additional public universities, new student services, and standardized admissions criteria. Some scholars have characterized this act as an early form of affirmative action for white working-class men, given its "active exclusion of African Americans and women and its abundant benefits for returning White male soldiers," which included the opportunity to attend college.[3] During the civil rights and Black Power movements of the 1960s, organized efforts pushed for the democratization of higher education through open admissions at public universities, a practice that limits enrollment criteria to factors such as a high school diploma. These students recognized the ways that structural racism was woven into the fabric of K–12 education, unfairly disadvantaging low-income students and students of color, who were far more likely to be educated in underfunded school districts with fewer resources and preparatory support structures for students.

The catastrophic rise of student debt in the United States is intrinsically tied to concerns that public universities were becoming hotbeds for leftist political organizing in the 1960s, especially in the California public college system. Ronald Reagan pushed to cut state funding for higher education and argued that a system of tuition payments and loans was both fiscally responsible and would help "clean up the mess" at campuses like Berkeley. By "mess," Reagan was referring to the militant student anti-war, anti-capitalist, and civil rights organizing of the period. Acting in concert with the FBI, and with the public support of prominent conservative intellectuals, Reagan successfully laid the groundwork for state funding cuts to higher education and the subsequent skyrocketing of student debt after the 2007–2009 Great Recession.[4]

Some universities created affirmative action programs in the 1960s and '70s as concessions to the demands of the civil rights movement and student activists of color. These programs were also meant to address vague federal guidelines following the Civil Rights Act of 1964 to remedy discrimination at federally funded institutions. Affirmative action for student recruitment allowed for special consideration to be given to prospective students from historically excluded groups. However, the 1978 Supreme Court case *Regents of the University of California v. Bakke* ruled that universities could not use racial quotas for admissions or use admissions to remedy societal discrimination. The case did recognize that a diverse student body was a compelling state interest and that race could be considered in addition to an applicant's qualifications. From its inception, affirmative action has faced legal challenges from white and, later, Asian American students, despite studies indicating that affirmative action policies do not harm the chances of students from these groups.[5]

In June 2023, the US Supreme Court significantly gutted race-conscious admissions programs at colleges and universities across the country, with military academies—institutions that aggressively recruit low-income students and students of color—exempted from the ruling.[6] Prior to this, however, many states had passed affirmative action bans by ballot proposition, state legislature, or executive order, including California, Washington, Michigan, Nebraska, Arizona, Oklahoma, New Hampshire, and Florida, resulting in significant drops in Black and Latine enrollment. According to education policy scholar Dominique Baker's 2019 study, between 1995 and 2012, as the proportion of white students at public flagship universities declined and concerns about the scarcity of educational resources intensified, the likelihood that a state would impose an affirmative action ban increased.[7] To remain in compliance with federal and state limits on affirmative action and avoid lawsuits, some universities have eliminated scholarships that specifically target racial minorities and women or have opened up scholarships and programs that targeted underrepresented students to white students.[8]

Efforts to increase "diversity," rather than explicit goals of desegregation and antidiscrimination, typically shape college recruitment practices among universities that consider race in admissions. Furthermore, recommendations from the US Department of Education's Office for Civil Rights and the US Department of Justice's Civil Rights Division after the 2023 Supreme Court ruling stress that colleges can continue to collect data about race and recruit students from schools that serve majority low-income students and students of color in the interests of promoting racial diversity on campus.[9] Yet critical scholars have pointedly argued that "diversity" helps embed US managerial discourse into higher education, often working to prioritize the preparation of predominantly white students for a globalized economy.[10] According to sociologist and oral historian Amaka Okechukwu, diversity "does not offer new visions for more equitable power relations or a radical reenvisioning of higher education. And yet conservatives' successful challenges to affirmative action and open admissions constrain the racial landscape through which we understand university administrators' efforts to make their somewhat progressive attempts to defend diversity appear even radical."[11] To explore Okechukwu's argument further in the context of the rise of smart universities, it is worth considering university decisions to no longer require standardized test scores for admission and how digital technologies are being deployed to measure and track prospective students for recruitment in light of this shift.

Smart Recruitment and Admissions

The SATs originally emerged out of testing developments at the turn of the twentieth century, when twelve college presidents from primarily elite universities in the Northeast founded the College Entrance Exam Board to preserve the racial exclusivity of their institutions in the wake of increased immigration during the Jim Crow era. As sociologist of higher education W. Carson Byrd explains, "Universities, particularly those associated with the Ivy League, sought testing approaches that supported narrow definitions of intelligence and merit while applying versions of 'geographic affirmative action' for white men from outside the legacy families and communities from which these institutions pulled most of their students."[12] The College Boards were administered for the first time in 1901, and for a fee, students could receive the subject matter beforehand.

After the development of the IQ test in 1905, the College Board test evolved into the SAT with the guidance of Carl Brigham, who had administered the army's experimental intelligence tests prior to joining the College Board test development committee. Proponents of the SAT argued that the exam was an objective, standardized instrument for predicting intelligence and future academic performance, thus providing a fair, quantitative metric for comparing different prospective students for admissions decisions. Many universities used the SAT's purported objectivity to provide justification for racist admissions policies. Brigham himself was a proponent of the idea that "racial mixing was degrading American education" and an active participant in the eugenics movement.[13]

In the aftermath of the Civil Rights Act of 1964, the SATs underwent public scrutiny for bias against poor students and students of color. However, the rise of the test-preparation industry further enabled affluent students to gain advantages in an exam that already privileges the types of social and cultural capital acquired through well-funded school districts.[14] Additionally, the College Board, which now boasts a revenue of $1 billion per year, has enabled college recruiters to license the scores, names, ethnicities, and family college histories of test takers who consent to having their information shared in the hopes of increasing their chances of admission. In practice, this information is used to create not only lists of students the universities hope to admit but also lists of students whom universities intend to recruit but not admit in order to inflate admission statistics to seem more selective and improve their rankings.[15]

A range of universities have recently moved away from requiring SAT and ACT scores for college admissions. Some have made this change because of the equity concerns scholars have been raising for decades and the proven feasibility of such a shift during COVID-19, when testing was suspended. Smart university recruitment initiatives are now promising to make up for these lost data points in an increasingly competitive higher education marketplace with a shrinking pool of potential applicants, given declining enrollments. For instance, the higher education marketing and recruitment company Capture positions itself as a solution to the problem that recruitment specialists will no longer be able to buy as many names from the College Board. According to one Capture representative, recruiters need to maximize the data that they *do* have access to by getting "smarter" through the use of machine learning to forecast, model, and nudge the "right" students to enroll.[16] Machine learning is the use of algorithms—a series of instructions provided to a computer—to provide predictions or make decisions based on training from large datasets.

A variety of universities use tracking software like Capture to monitor prospective students' engagement with university emails and websites, to match anonymous visitor data with the information prospective applicants share with universities, and to create targeted digital marketing based on prospective student information and behavior, including email addresses, geolocation, browsing history, and social media engagement. Although universities typically disclose that they use cookies on their websites—meaning data generated by their web server and stored on prospective students' browsers to collect their information and preferences, identify and remember their devices, and track their online activity—it is not always clear how prospective students are being tracked or how they can opt out. These tools are protected from direct scrutiny as the proprietary technology of the institution and its private partners.

Data-informed outreach strategies, like Capture's, are marketed as helping university recruiters provide personalized and relevant automated messaging that will trigger applications and enrollment.[17] Furthermore, Capture offers a dashboard recruiters can use to manage yield and hit enrollment targets, which includes, according to their website, "a detailed profile of all your prospective students, including an Engagement Score and an Affinity Index" that will help recruiters "direct time, resources, and dollars to where they will have the most impact."[18] Although dashboards for higher education admissions are often marketed as offering greater transparency and control to their users, in practice, dashboards often do not provide insight into the underlying algorithms and data-selection processes that support the specific user interactions the dashboard enables for higher education decision makers.[19]

There are also significant concerns that tracking software and scores for prospective students based on their behavioral data, such as the ones Capture offers, will contribute to recruitment processes that already disadvantage marginalized students. For instance, there is an assumption baked into many higher education digital tracking efforts that student engagement with a university website is indicative of their level of interest and likelihood to enroll if admitted. Yet for rural and low-income students who do not have reliable access to the web, their interest is not likely to be reflected through website visits.

Furthermore, the use of historical data from university enrollments to train predictive tools is also alarming, given that existing enrollment data in higher education disproportionately favors white and Asian students, who are more likely to attend and complete degrees at four-year institutions.[20] Capture's proprietary software goes so far as to forecast "the likelihood of students from a specific ZIP code applying to your school, identifying clusters of students who match specific search criteria."[21] This use of zip codes is concerning, given the ways that long-standing histories of racial and class segregation pattern where people live in the United States, which subsequently shapes distributions of resources and life chances.

People from majority-white zip codes attend college at the highest rates, are more likely to graduate, and take on less debt to do so. This is due to the structural advantages afforded by well-resourced school districts and the accumulation of intergenerational wealth, in contrast to racial groups who have experienced discrimination, criminalization, and wealth expropriation for generations.[22] As universities invest resources in software companies that promise to identify prospective students who are most likely to enroll, be retained, graduate, and become alumni donors in the future, it is hard to imagine how notions of a "right fit" student won't be biased to favor those who are structurally advantaged. Instead of investing these resources in equity-minded, robust recruitment and support services for minoritized students—which would require, for many institutions, organizational change—universities are instead trying to use technology to admit individual students best positioned to "succeed" in the neoliberal academy. Additionally, although some promoters of digital tracking technologies for recruitment mention the possibility of using these tools to identify and target underrepresented groups in the academy, such as first-generation students who are Pell Grant eligible, these grants have not kept pace with the rapidly increasing costs of a college degree, which means these students are likely to take on detrimental student loan debt if they are successfully targeted by the institution.

According to ProPublica's analysis of data from the US Department of Education, public colleges and universities have been giving a declining number of grants to students in the lowest quartile of family income, despite the fact that tuition has increased dramatically at many public institutions.[23] Over the past two decades, the nation's lowest-income students are increasingly being educated by community colleges and for-profit institutions with far less support and lower graduation rates, rather than four-year state schools. It is unlikely that services like Capture will reverse these trends, given that it provides services that are optimized according to dominant understandings of institutional success metrics under austerity.

One example of this approach for optimizing recruitment budgets is universities' use of financial aid leveraging strategies to entice applicants targeted to raise the institution's rankings or revenue streams. Rather than offering substantial aid to those most in need, companies such as CampusLogic promise to help higher education administrators "leverage micro-scholarships to fill your classes with better-prepared students" through the use of digital tracking and personalization tools.[24] Similarly, Capture's AID tool uses machine learning and admitted student data, including deposit data, contextual neighborhood-level data, and behavioral data, to offer "aid to the individual students with the most impact."[25] Impact, here, refers to the optimization of financial aid spending for university recruitment goals. According to Capture's website, their dashboard allows users to "easily integrate, adjust, and download student and group-level probabilities and aid adjustments" to "target discretionary funds where it will have the most impact on enrollment for the least additional spend."[26] Although universities participating in federal student-aid programs are legally required to disclose the criteria they use to select aid recipients, existing regulations do not dictate what level of detail institutions need to provide.

The data underpinning services like Capture is not independent of the historical, ideological, and economic conditions that give rise to that data's production, capture, and use. In US higher education, most institutions are primarily funded by tuition and then gifts, and compensation and aid are some of the largest expenditures. The intersection of market pressures and funding cuts with universities' commitments to diversity rather than social justice and systemic reform has led many public institutions to prioritize rankings, revenue generation, and funding performance metrics rather than offering substantial support to prospective low-income and minoritized students who are most in need of aid. Software designed to predict which students are likely to enroll, and under what conditions of admission, is a way of further consolidating power in the hands of the predictor (universities in partnership with private software companies), rather than the subjects of that prediction (students). Additionally, Capture's emphasis on offering budget-conscious solutions that maximize tuition dollars and help recruiters demonstrate "impressive return on investment" to university presidents reflects the institutional conditions of austerity in higher education, which then shapes the data points Capture prioritizes and the types of power they seek to shore up for university recruiters.[27]

Capture's marketing team, like many other smart university initiatives, appeals to prospective consumers of its higher education products by drawing parallels to Big Tech. According to Capture's senior enrollment strategist, higher education has been slow to adopt the digital marketing strategies of companies like Amazon and Google, to its detriment. In one of Capture's promotional Webinars, Amazon's use of machine learning, in particular, is celebrated for its ability to offer tailored recommendations based on tracking people's purchasing habits and preferences. There is no engagement with the question of how machine learning, be it in the context of Amazon's recommendation algorithm (discussed in chapter 1) or targeted higher education recruitment, might raise legitimate concerns about discrimination.[28]

Machine learning uses algorithms to identify, analyze, and draw inferences from patterns in data. Common applications of machine learning include Netflix and Amazon recommendations and digital assistants like Siri. Machine learning is what makes it possible for these digital technologies to detect patterns from user information to predict and nudge their behavior in a range of ways. Additionally, machine learning is rife with issues of discrimination and unequal distributions of power over who gets to design these tools and who tends to be most adversely impacted by them. For instance, in the criminal legal system, machine learning tools like Correctional Offender Management Profiling for Alternative Sanctions (COMPAS) are used to assign risk scores to criminal defendants for informing bail and sentencing decisions.

COMPAS disproportionately and falsely labels Black US criminal defendants as high risk for committing future crimes because of the ways its training data reflects past patterns of historical discrimination. Furthermore, the tool itself is fundamentally biased toward incarceration. Nonpunitive solutions such as ending pretrial detention and cash bail have not been measured in a manner conducive to machine learning, nor do they map as neatly onto dominant cultural assumptions and legal norms that uphold mass incarceration in the United States.[29] The oppressive use of machine learning within the criminal legal system and the university's role in supporting its application in this domain is discussed extensively in chapter 5.

When we consider examples of machine learning's detrimental effects on marginalized people's life chances, it is important to understand that bias and discrimination are not merely computational problems but rather social ones with root causes. It is not simply that the datasets used to train machine learning emerge from, and thus often reflect and reproduce, patterns of inequity in society but also that the fundamental act of using machine learning to reduce, formalize, and gather feedback is itself not a neutral or objective process.[30] Those who build and finance the development of machine learning tools embed their goals and beliefs about what these automated technologies should optimize for and how. Even if it were possible to ensure that machine learning and other tools were being used to support and not undermine the recruitment of marginalized students, such an approach still sidesteps the fundamental inequities plaguing higher education, including but not limited to the disparate consequences of rising student debt and the lack of support structures for minoritized and nontraditional students after admission. The recent spate of legislation to dismantle diversity, equity, and inclusion programs and initiatives across several US states is also poised to undermine the limited forms of support that do exist for marginalized students at public universities. By limited, I am referring to both the underfunding and institutional marginalization of DEI efforts, as well as the limits of DEI efforts themselves, which are often divorced from addressing deeper systemic injustices.[31]

The history of higher education admissions also teaches us that the admissions process often perpetuates racial inequality in a multitude of ways, and not simply through the outputs of machine learning algorithms, from bias in the criteria used to evaluate prospective students, to the abandonment of affirmative action, to student interactions with recruitment professionals. For instance, sociologist Ted Thornhill's audit study of emails sent to white college admissions counselors from fictitious Black students found that counselors would not respond to students "too concerned" with race or racism or otherwise discourage them from applying.[32] Instead of having higher education funding and admissions models equipped to address deeply entrenched, long-standing patterns of discrimination, universities are turning to a range of digital tools to maximize recruitment dollars in ways that can further discrimination against racially and economically marginalized people.

Social Media, Recruitment, and the Noncitizen Student

In addition to more automated forms of targeting by tracking students' online behavior, including their social media engagement, college admissions committees have also revoked admissions offers based on students' social media posts.[33] According to communication scholars Brooke Erin Duffy and Ngai Keung Chan, parents and educators have increasingly begun socializing young people to anticipate the imagined gaze of colleges and future employers, another form of big data socialization. These scholars argue that "a hidden curriculum that prods individuals to expect scrutinization of their personal social media profiles is the same one that renders professional surveillance permissible ... [I]n workplace surveillance—much like social media—the boundaries between the personal and professional get blurred in alarming ways."[34] Although this normalization of institutional monitoring is alarming, it is also important to note that social media surveillance does not impact students across social groups evenly. Social media surveillance builds on a history of disparate surveillance for noncitizen students.

Following September 11, 2001, in particular, international students became increasingly subject to harsh US government visa policies, including the enactment of the National Security Entry-Exit Registration System and the Student and Exchange Visitor Information System. The former required immigrant men from predominantly Arab- and Muslim-majority countries to submit biometric data directly to immigration officials, and the latter all visa applicants to interview with consulates and submit to biometric screening in their home countries.[35] However, during and after the 2008 financial crisis, and in response to declines in international students and the inability of domestic students to pay full tuition, US universities began targeting international students for recruitment as a means of generating additional tuition revenue. These efforts have been coupled with intensifying forms of surveillance for international students, and in particular, students racialized as nonwhite and traveling from countries that the US government frames as "threats" to national security.[36]

Prospective international students often encounter surveillance and discrimination through delays and denials of the visa provision process in their home countries and through denial of entry into the United States even after having obtained a visa. Once on campus, as feminist studies scholar Abigail Boggs explains, these students experience "the surveillance, policing, and precarity of their lives and educations, since maintaining a visa requires continuous approvals from and registration with international student advisers who have been effectively deputized as immigration officials."[37] Even after a student has been granted admission to a US university and approved for entry by the State Department, US Customs and Border Protection can use its discretionary power to deny international students entry into the United States. This unjust denial of entry was recently reported in the *New York Times* as happening to Iranian students.[38] There are reports of immigration officials using international students' social media as grounds to revoke visas, including a Palestinian student living in Lebanon who, after a five-hour interrogation in 2019, was "deemed inadmissible to the United States" owing to his friends' political posts on social media.[39] Although this decision was subsequently reversed, it nonetheless captures the ways prospective international students have had their whereabouts, political views, and personal information monitored by both universities and the US government in ways that significantly hinder their freedom of expression. International students have also been threatened with deportation if they participate in unionization efforts at their institutions, including at the University of California, Santa Cruz, and Washington University in St. Louis. Yet some software development companies argue that it's precisely international students—particularly from Hong Kong, China, Singapore, and the UK—who can be attracted to universities that embrace a "connected learning experience."[40]

Thus, when universities claim to be incorporating equity considerations into their recruitment processes, it is important to ask whether the academy these students are being admitted into is actually ameliorating, or instead reproducing, structural inequalities baked into higher education. As this chapter has demonstrated, most universities were not founded to be race neutral or meritocratic but, rather, have generally supported racial stratification and inequality, even in historical periods where universities have deployed purportedly objective or neutral admissions criteria. Presently, having a four-year degree significantly impacts lifetime earnings, although earning potential with a college degree is higher for white men than for other demographic groups. Additionally, African Americans of all genders are also more likely to enroll in for-profit colleges than their white peers, which endow these students with comparatively more debt and hold less value on the job market.[41] At both public and for-profit colleges, students of color are more likely to have to rely on debt to pay for college, to have to borrow more than white peers, and to face more barriers in their efforts to repay. On average, Black degree holders' debts grow after graduation, in contrast to other demographic groups, because of income inequality in the workforce. When looking at the intersection of race and gender, Black women incur the highest amount of student loan debt on average.[42] Ultimately, rather than offering meaningful forms of redress for racial injustice in higher education admissions, smart recruitment initiatives might help preserve majority-white institutions or modestly diversify these universities while shoring up higher education's move toward privatization, financialization, and industry capture, with disparate consequences for low-income students, noncitizen students, and students of color. The following chapter looks at how digital monitoring is transforming student retention efforts, with particular attention to their implications for marginalized students labeled "at risk" for dropping out.

Chapter Three

Retention

According to the rationales underpinning smart university efforts, by rendering students knowable through the use of digital technology, their learning experience, well-being, and progress to degree can purportedly be predicted and guided. Although all students become objects of inquiry for university surveillance during their time in higher education, the intent and impact of that surveillance plays out unevenly for marginalized students.[1] As just one example, at primarily white institutions, Division I Black male athletes tend to face myriad forms of hyper-surveillance, including the watchful gaze of athletics personnel, restrictive policies governing their use of social media, and the hiring of class checkers to document whether these students arrive to class and remain for the full class period—a practice that is becoming increasingly digital and automated. In addition to procuring digital monitoring services that flag posts with potentially "offensive" content, some colleges require student athletes to provide the passwords to their social media accounts. Others require students to participate in workshops outlining unacceptable social media behavior.[2] While there have been many high-profile examples of student athletes using social media to engage in racist and homophobic behavior, there are others who use their platforms to challenge injustice on their campuses and in their communities. Student athletes are in some sense more likely to become embroiled in public relations incidents for universities compared to others because of their hyper-visibility on social media platforms. They are also some of the most regulated student populations owing to National Collegiate Athletic Association (NCAA) bylaws that stipulate rules regarding curfews, physical examinations, and academic performance, among others.[3]

Black student athletes often learn within hostile campus racial climates where deeply rooted racial stereotypes shape their interactions with professors, peers, and coaches, and where their eligibility to compete in sports is prioritized over their intellectual development. Their relationships with universities are bound by contracts that typically tie their financial aid to their participation in the athletics program. As education scholar Eddie Comeaux explains, "Poor graduation rates among Black male athletes are indicative of the systemic failure—and role of race and racism—in which they are largely celebrated and economically exploited for their athletic prowess; yet, they are seen as anti-intellectual and even marked as disposable once their athletic eligibility ends."[4] Even as institutions rely on increasingly technologically mediated and inflexible forms of classroom attendance tracking for student athletes, it is the scheduling demands of coaches and television networks that often cause student athletes to miss class. In some cases, Black male athletes are restricted to certain majors to limit scheduling conflicts. Comeaux describes these hyper-surveillance practices as motivated by neoliberal governing structures and as drivers of structural racism, exploiting Black male athletic labor while failing to ensure the athlete's academic experience is fair and just.[5] Student athletes have been fighting for the right to be recognized as employees under the Fair Labor Standards Act (FLSA), most recently in *Johnson v. NCAA*. FLSA would require student athletes to be paid in revenue-generating sports akin to work-study classmates. One of the NCAA's core legal defense tactics to justify the exploitation of student athlete labor hinges on the slavery loophole of the Thirteenth Amendment, which permits forced prison labor.[6]

Athletics stakeholders often place blame on individual Black male athletes for academic underperformance as opposed to institutional failures, and this blame is used to justify the surveillance of their choices and behaviors. As information studies scholar Philip Doty explains, student athletes' "study habits, use of library resources and other academic materials, use of libraries' presence on social media, libraries' Wi-Fi networks, library buildings, and so on, receive greater scrutiny than the behavior of other students. Thus, they are stigmatized externally as potentially incapable of success and autonomy, whether academic or otherwise."[7] While student athletes are arguably some of the most regulated and scrutinized students on campus, many smart university initiatives for student retention operate under a similar premise: that digital data can be used to identify which students are "at risk" for underperforming or not passing a course, disengaging from campus life, dropping out, or needing additional guidance.

This chapter demonstrates how notions of risk get constructed through universities' development and use of digital technologies for student retention. In particular, it emphasizes how this idea of risk supports monitoring, sorting, and classifying students in ways that can be punitive and stigmatizing, particularly for already marginalized students. First, the chapter briefly historicizes the idea of risk that retention technologies help materialize and its relationship to race. It then details how the present-day integration of digital tools for student retention, often done in partnership with for-profit tech companies, can raise issues of student privacy, discrimination, and exploitation, as well as undermine academic freedom.

Perhaps most significantly, digital tracking technologies for retention often locate the problem of student performance in their behaviors, rather than in institutional and societal failures to adequately support students.

Unpacking Risk

Although a growing body of scholarship locates risk within the history of neoliberalism and the proliferation of audit cultures, citation metrics, and accreditation agencies, the notion of risk can also be situated within a longer history of racial oppression the United States. According to anthropologist of computing and mathematics Rodrigo Ochigame, throughout the nineteenth century in the United States, the concept was tied to corporate risk management, largely concerning "antebellum legal disputes over marine insurance liability for slave revolts in the Atlantic Ocean."[8] With the rise of modern mathematical statistics, US postwar neoliberal-era market institutions increasingly used actuarial techniques to sort, classify, and control the life chances of populations.[9] By the second half of the twentieth century, actuarial methods were a key feature in policing, parole decisions, and credit bureaus. These methods were often couched in promises of transparency and equal, dispassionate treatment. When 1970s US civil rights and feminist activists challenged the use of risk classification in the pricing of insurance as unfair and discriminatory, the insurance industry promoted the idea of actuarial fairness as a complex, technical, objective, and apolitical matter in order to evade regulation.[10] Similarly, the present-day use of risk classification for predictive policing is marketed as a more neutral and objective means of distributing police to communities, despite the fact that these tools perpetuate and amplify the surveillance and mass incarceration of people of color.[11]

An analysis of smart university initiatives for student retention helps demonstrate how present-day ideas about at-risk students are informed by historical notions of risk that are racialized and carceral. The rationale underpinning risk mitigation for digital retention tools is not focused on the economic and social risk that students, and especially marginalized students, have experienced from market deregulation, exorbitant tuition fees, and discriminatory institutional practices, but rather risks associated with attendance data and frequency of visits to the library or other campus services gathered from Bluetooth beacons and WiFi access points. At Temple University, their early-alert system for identifying at-risk students was designed, in part, by a former criminologist who helped build predictive models for analyzing the probability that a given criminal defendant would reoffend.[12] Such technologies are notoriously skewed against Black defendants and yet continue to be used in the US criminal legal system. Furthermore, similar to the ways that risk mitigation tools for student retention sidestep the structural conditions that shape student behavior and performance, risk assessment algorithms in the US penal system do not engage with how the institution of policing itself contributes to the conditions that produce crime, including through the criminalization of poverty and the hyper-policing of low-income communities of color.[13]

In the context of recruitment in higher education, once a student is labeled at risk, the onus is typically on the student to better their choices and behaviors in response to digital or advisor nudges, while the technology collecting their data is treated as an objective institutional process. In some cases, the consequences can be quite punitive, such as revoking scholarship funds based on attendance data.[14] Furthermore, there are other connections to the carceral state underpinning smart university initiatives for retention: several cloud-based learning management systems, like D2L's Brightspace, rely on Amazon Web Services for hosting. Universities are thus sending data to Amazon at the same time that Amazon workers and activists are pushing back against the company's exploitative logistics empire, surveillance partnerships with local police departments and Immigration and Customs Enforcement (ICE), and retaliation against Black warehouse workers who organized against company practices during COVID-19.[15]

Present-Day Digital Tools for Retention

Digital technologies for retention involve collecting student data to tailor the teaching, learning, or advising experience, as well as to nudge students toward preferred patterns of decision-making. Such tools include learning analytics to target curriculum, instruction, or advisor support; LMSs and dashboards that offer data on when, where, and how students engage with online learning platforms or their overall course performance relative to other students; e-advisors that send automated nudges to students to guide course enrollment and performance; and beacons, routers, sensors, and mobile devices for tracking student attendance and utilization of university resources. Under the smart university, student data is not only used to shape student decision making and progress to degree but also to market the university to prospective students, to serve as business intelligence for higher education governance, and for policymaker and politician inspection when it comes to evaluating institutional effectiveness and whether to withhold funding.[16]

Students rarely have knowledge about whether, how, or where, precisely, they are being tracked. Some companies, such as SpotterEDU, actively conceal the beacon technology underpinning their services so that it cannot be gamed or sabotaged. In one reported case at the University of North

Carolina, the university installed SpotterEDU for tracking student attendance without notifying faculty or students, causing alarm and confusion on campus.[17] Several students have also claimed that SpotterEDU marks them absent or late even when they are on time and present. The efficacy of learning analytics for improved student learning has been called into question by researchers from a range of disciplinary backgrounds, and ethical issues are rarely engaged with in learning analytics research.[18] Ultimately, these tools shore up power for the technocratic management of students' progress to degree in order to meet educational performance metrics and shift power over teaching and learning conditions away from instructors.

While the collection of student data for planning, operational, and pedagogical purposes is not new, shifts toward "smart" digital tools provide unprecedented levels of granularity. Companies like Nestor AI and Corerain are now offering emotion recognition technology for measuring student attentiveness, and some universities like Arizona State are experimenting with emotion and facial recognition software to theoretically discern whether students are confused, attentive, or enjoying their time.[19] Efforts to innovate digital tools that focus on improving individual student performance through surveillance and behavioral control persist despite mounting evidence that circumstances largely *outside* of students' individual control—their finances, campus climate, and other forms of structural inequity that get flattened into "demographic" traits—are the greatest predictors of students leaving the academy.[20]

This flattening effect is compounded by the fact that researchers have consistently demonstrated racial bias in both facial and emotion recognition technology.[21] Furthermore, as communication scholar Sun-ha Hong reminds us, "When ostensibly superior data analytics enter into a social problem, they tend to de-value and invisibilize the kind of data that cannot be captured easily and the kind of judgement that cannot be rationalized through the available data."[22] When digital tools for retention are integrated into the decision-making practices of faculty, advisors, and higher education administrators, they structure how the problem of student retention is understood and what seems most possible to do to address it. This sidelines frameworks and approaches that are un- or underfunded, more time intensive, or not conducive to a technical fix.

As just one example, in 2009, Purdue University introduced one of the earliest student-tracking initiatives for retention using predictive analytics, known as Signals. Using student data including grades, test scores, dining-hall use, and other data points, Signals worked to determine which students were on time to graduate, assigning them risk colors and offering a system of nudges to instructors for pushing emails, reminders, and text messages. However, critics questioned whether the tool actually increased retention as claimed, and the program was eventually stopped.[23] A similar program launched at Georgia State University has also been using predictive analytics to guide student decisions about their major based on how their data compares to past students who have successfully graduated. While the university's graduation rates improved, it was at the expense of directing students of color to fields with less earning potential.[24]

Private companies have largely come to replace university-produced platforms for retention management, and public universities have been eager adopters of these services in the hopes of improving retention and graduation rates. Yet the use of predictive analytics for retention has gone hand in hand with many public universities doing away with remedial education for underprepared students, increasing student-faculty ratios, and increasing reliance on non–tenure track faculty, all of which have been found to negatively impact retention and graduation rates.[25] As the ability for faculty and students to develop close bonds is increasingly strained, data, and the tracking and predicting of student behaviors, emotions, and actions, is proffered as a solution. Rather than reversing these trends, universities are turning to digital tools that promise "technology-fueled efficiency and precision" for managing student retention, discursively aligning higher education with the technological solutionism of Silicon Valley.[26] Technological solutionism refers to the reduction of complex social phenomena to neatly defined problems that can be solved and optimized for using technology.[27]

Proponents of retention tools at public universities, such as the tracking of student dining-card swipes to measure "social connectedness" with other students, argue that big data can help "level the playing field" with private universities, which are able to offer small student-to-faculty ratios and high levels of personal attention.[28] It's important to acknowledge that with surveillance, control and care are not always easily distinguishable and that benevolence and coercion can coexist.[29] Yet this mode of surveillance as a means of measuring students' sense of belonging simultaneously accepts as a given the conditions of bloating class sizes at many public universities. Additionally, public universities have been some of the swiftest adopters of smart retention tools given that they enroll more lower-income and first-generation students than pricier private universities.[30] This deficit framing—that the academic performance of marginalized students is a problem to be solved through surveillance technology—helps shift critical attention away from the institutional structures failing students and toward questions of how to best track, measure, and nudge student performance. It's also the case according to New America senior advisor Iris Palmer that being identified as at risk can "encourage a self-fulfilling prophecy, or create a stereotype threat ... First-generation students, especially, may interpret an attempt at offering help as a sign that they're not college material."[31] Some companies have emerged that offer data analytics and business intelligence specifically for historically Black colleges and universities, including for tracking student retention, which disproportionately serve first-generation students. Such tools promise to help optimize financial operations while improving student outcomes in a higher education context where HBCUs are systematically underfunded and under-resourced.[32]

University shifts toward privatized technology have also included LMSs for higher education like Blackboard, Canvas, Brightspace, and D2L, which run on private cloud infrastructures. This not only makes it harder to audit platforms for privacy compliance and to receive meaningful informed consent from users, but it also raises issues of academic freedom. As computer security researcher Tobias Fiebig and collaborators argue, "If, for example, a U.S.-based LMS provider decides to enforce U.S. sanctions against citizens of specific countries for an LMS, including customers outside the U.S., it can effectively dictate which students a university enrolls by controlling the 'means of study.'"[33] This scenario is not far fetched if one considers the 2020 case of Zoom preventing faculty and students at New York University from holding a guest lecture with Palestinian activist Leila Khaled, arguing that it violated the terms of their license, or GitHub, a platform for software developers to store and manage code, restricting accounts to comply with US trade sanctions.[34]

Brightspace's web presence emphasizes its ability to "help students, faculty, and advisors all make more informed and timely decisions," arguing that all university members benefit from the platform's tools and services. At the same time, within this discourse, while institutions and advisors are "empowered" to find the data points they need, students are "engaged" as objects of inquiry for tracking regarding academic and persistence risk.[35] A close reading of Brightspace's own marketing rhetoric reveals the power dynamics laden in the system's design and functions, despite the presentation of the software as an "equitable, all-encompassing ecosystem" where its benefits are evenly distributed.[36]

Major cloud providers use lobbying, international initiatives, and economic incentives, including the promise of scalability and support for university storage and communication needs, to become default infrastructure for higher education. It is perhaps unsurprising that the use of cloud technology in universities has increased most significantly in countries where administrators and business managers oversee university operations and where public higher education has shifted decisively toward privatization.[37] Furthermore, the LMS is only one piece of the global higher education technology market, which is projected to grow to a value of $370–$410 billion by 2025, based on the current explosion of edtech deals and an influx of venture capital investment into edtech companies.[38] Established foundations like the Hewlett and Gates Foundations have also been granting upwards of $20 million to higher education initiatives, including $3 million toward learning analytics projects as part of their Next Generation Learning Challenges program.[39] While some scholars locate the emergence of learning analytics in increased university investments in online distance learning in the mid-2000s, integration of advanced technologies like natural language processing and machine learning into higher education is growing, particularly in the wake of the COVID-19 global crisis. Furthermore, the sense of crisis surrounding the pandemic and the need for a rapid shift to widespread online learning gave edtech businesses increased opportunities to sell untested solutions that have little to no connection to trusted teaching and learning philosophies.[40]

Given the lack of federal privacy laws governing student data brokers, student information can be collected, sold, bought, and leveraged for improving private companies' products and services. The privacy policies of major LMS providers like Canvas are often vague as to whether the data can be sold to third parties or shared for advertising purposes. Additionally, the private equity firm Thoma Bravo purchased the educational software firm that owns Canvas, prompting a public letter urging the firm not to abuse its access to student data.[41] The University of California (UC) committed to investing $200 million in Thoma Bravo after using Canvas to repress graduate student workers fighting for a cost of living adjustment at UC Santa Cruz, including through the use of IT experts to retrieve grade deletion data to punish striking students. Ultimately, smart university initiatives for retention raise considerable concerns about student privacy, discrimination, and the exploitation of student labor for the refinement of digital products and services. The following sections look at each of these concerns closely using concrete examples of retention tools used in public universities across the United States.

Student Privacy

Many smart university initiatives involve third parties, beyond the university and the immediate vendor of the digital platform or service. For instance, LMSs offer plug-ins that can introduce additional parties with their own privacy policies and terms of service. While university contracts with vendors are inadequate for ensuring fair, democratic distributions of power when it comes to decisions about whether and how to incorporate digital tools onto college campuses, contracts nonetheless provide some structure for transparency and negotiation. Yet partnerships between vendors can violate even the most baseline forms of student data policy in higher education. A range of student data points, including educational background, academic interests, and contact information, are often shared as leads multiple times for different vendors, who have their own terms of use.[42]

Tools like LMSs also give faculty and advisors unprecedented levels of control for monitoring students, including details about how much time they spend on the platform, where they log in from, and at what time they complete assigned work. Attendance trackers and the collection of GPS and card-swipe data can also have a chilling effect on students' capacities to collectively organize and exercise autonomy over their own learning.[43] However, both students and faculty can experience disempowering forms of surveillance from the use of digital tools for retention. For instance, information science scholars Kimberly Williamson and Rene Kizilcec found in focus groups that students and faculty expressed a lack of clarity regarding who can access the data taken from dashboards and what decisions would be made on its basis.[44] In addition to a loss of academic freedom in the classroom, instructors expressed concern about the extra work needed to keep up with proliferating dashboard related metrics.

Although the notion of privacy can help identify ethical issues with smart university initiatives, privacy concerns tend to crowd out a range of other important ethical considerations. Among academic and education policy leaders, privacy is often understood as compliance with standard fair information practices and traditional ideas of informed consent and confidentiality.[45] More broadly, conversations about the ethics of information technology in the United States have generally been framed in terms of privacy at the expense of issues regarding racial discrimination, autonomy, and economic exploitation. When ethical issues concerning datafication are framed in terms of privacy, the question is often how to ensure data is collected anonymously and stored securely and that students can readily opt out. However, we can also ask, should these tools be deployed at all? What values, goals, and assumptions concerning the purpose of higher education do they mask as purely objective or efficient forms of technical problem solving?

For example, it is not enough to ask how the venture capital–backed educational technology company EduNav collects student data and whether these practices are secure, anonymized, and transparently communicated to students. According to EduNav, this data is collected to purportedly increase student access, equity, and affordability by improving time to degree, optimizing course schedules, and streamlining the transfer process.[46] Additionally, EduNav's SmartPlan advises students regarding what courses to prioritize to "speed up graduation and minimize cost."[47] Such a framing makes students responsible for the political and economic pressures of austerity in higher education and treats these pressures as natural, given, and immutable. Students must speed up, approach their learning as instrumentally as possible ("stay on track"), and finish as quickly as they can in the midst of exorbitant tuition fees. E-advising efforts to guide students on what courses to take based on their chosen majors, graduation requirements, and academic performance relative to peers raises considerable concerns about reproducing inequity in a context of rapidly declining funding and decreasing enrollments in the humanities at many institutions.

Discrimination

There are also significant concerns that predictive analytics for student success will intensify discrimination. In anthropologist Madisson Whitman's ethnographic study of these tools, university data personnel describe their efforts to design models based on behavioral data that represents things "students can change" to nudge students to make certain choices that are correlated with retention. In contrast, things students "can't change," meaning demographic data on race, ethnicity, gender, parental income, and high school zip code, are considered "outside of the model's purview because while they correlate with graduating in four years, they are not actionable."[48] This seemingly race-, class-, and gender-neutral approach to predictive analytics conceals systemic inequity in several key ways. First, by presupposing that students have tremendous agency over their choices, this model erases the circumstances and structural constraints that differentially shape what choices are truly available to whom in the academy. This failure to engage with interlocking forces of oppression that shape student experiences—racism, sexism, ableism, class inequality—is not unique to predictive analytics for student success metrics but rather a common feature in dominant design practices.[49] Furthermore, Whitman's research demonstrates that even when confronted with inaccurate or incomplete data, data personnel adhere to the notion that the collected data are reliable proxies for student behaviors, and that student behaviors can and should be abstracted from demographic data for the purposes of modeling.

It's also important to note that demographic data is not representative of natural or objective forms of identity classification but, rather, political categories that have been constructed and contested over time. These categories are inseparable from the systems of stratification that unfairly order the distribution of opportunity and harm in society.[50] As sociologist Carson Byrd explains in the context of racial equity concerns in higher education, the "common association of inequality with individuals and not with the organizational reality of college campuses exemplifies how universities operate as racialized organizations; the organization is framed as race-neutral despite much evidence to the contrary."[51] The university's investments in surveillance technology to relentlessly gather, mine, and analyze student data shift focus away from institutional failures to redress oppression within the higher education system, and toward student behaviors instead. Little consideration has been given to how students are impacted by messages that they need additional oversight, intervention, and surveillance. These impacts will weigh differently on marginalized students who are disproportionately labeled at risk.[52]

Furthermore, many of the surveillance infrastructures used in the name of student success double as ways of monitoring international students' compliance with enrollment, attendance, and academic conduct requirements, which is then pooled with data from the Student and Exchange Visitor Information System. Through these digital infrastructures, educational providers essentially become surveillance workers for enforcing immigration laws and restrictions. The level of granularity afforded by these technologies also means that minor administrative violations and infractions are more easily flagged for early intervention to anticipate, prevent, or capture "various crimes of mobility."[53] This same group of students has less entitlement to financial aid among other social services as well. Thus, smart university initiatives for retention can help enforce extractive and coercive forms of governance over noncitizen students, the majority of whom arrive from non-Western countries.

Exploitation of Student Labor

A far less discussed implication of smart university initiatives for recruitment is that they are predicated on the exploitation of student labor. This labor, as privacy scholar Chris Gilliard points out, is in many cases "rendered invisible (and uncompensated), and student consent is not taken into account. In other words, students often provide the raw material by which ed tech is developed, improved, and instituted, but their agency is for the most part not an issue. It should be."[54] The exploitation of student data not only impacts the student providing that information but also other students, given the ways student data inputs are aggregated for predictive analytics. Student data typically becomes an indefinite asset of universities and private firms once collected, especially once deidentified.[55] There is also a sense of entitlement to student data not only among university administrators and private education technology firms but in many cases, among university researchers who contribute to the development of their campus's tools. The types of monitoring technology used to gather, predict, and control student behavior can also be developed and deployed for other contexts, from retail and workplace management to crowd control and potentially other forms of repressive workplace and urban governance.[56]

There are some digital tools for retention that do allow students to opt out, and in cases like these, proponents are quick to point out when few students exercise the option to do so.[57] However, utilizing a system that is opt out as opposed to opt in puts the onus on students who must then choose whether to access the services that institutions say are key to their academic success. It is not enough to ask whether learning analytics and other tools benefit both students and institutions equally or fairly (although part of this chapter's argument has stressed that these benefits are asymmetrical). These tools help shape distributions of power in the academy toward black-boxed platforms and away from students and faculty.

More fundamentally, this chapter has argued that many of the digital tools used for retention prioritize the meeting of student performance metrics in ways that foreclose other types of solutions that might not lend themselves as easily to metrification. This subsequently transforms what counts as student success and who should be the arbiter of when that success has been actualized. Currently, technical solutions are being largely privileged over solutions that require pushing back against the idea that public institutions can't afford to decrease student-faculty ratios or to increase academic support services to redress long-standing systemic inequities in the US education system. Although in theory, increased transparency about how these digital tools are being funded, designed, and deployed in higher education would potentially help efforts to organize against these measures, transparency does not necessarily equate to accountability or democratic control over whether a given tool should be used at all. While transparency is important, it is certainly not enough. An emphasis on transparency can also sidestep the question of whether continuous surveillance and privatization in higher education should be acceptable in the first place. In the next chapter, I look at how student mental health intersects with issues of surveillance and privatization, and how smart university initiatives build on preexisting individualist notions and practices of student wellness.

Chapter Four

Wellness

Mental health and student wellness are increasingly occupying a central role in discussions about the future of higher education. Colleges across the United States are reporting increases in the prevalence and severity of mental health conditions.[1] A 2019 study of more than two thousand graduate students in twenty-six countries found that PhD students suffer from anxiety and depression at rates that far exceed the general population.[2] Both graduate and undergraduate students from racial, ethnic, gender, and sexual minority groups experience disproportionately higher rates of poor mental health outcomes due to factors that include cultural stigma, discrimination, and a lack of adequate access to resources and structures of support.[3] This chapter focuses on universities' adoption of WellTrack as a solution to this mental health crisis.

WellTrack is a smartphone application modeled on cognitive behavioral therapy techniques. It is designed to help users assess and track their own mental health, learn about the causes of anxiety and depression, and utilize the application's self-help tools. WellTrack is a subsidiary of CyberPsyc Software Solutions, Inc., a private company based in Canada and founded in 2010, which aims to capitalize on the expanding market for anxiety and depression treatment in the North American economy. This market is valued at an estimated $46 billion each year. CyberPsyc's clients have included companies such as HCL America, as well as several public universities in the United

States and Canada, including Georgia State University; the University of Miami; Ohio University; the University of California, Santa Cruz; Purdue University; Ball State University; Montana State University; Ryerson University; York University; and the University of Alberta.

This chapter argues that WellTrack is one of the ways US public universities are investing in making their campuses smarter. The app promises to make university services more efficient and effective using student self-tracking to temper demand on university mental health services and by offering a real-time dashboard of aggregate student mental health data for administrators. First, the chapter situates the rise of the WellTrack app within the history of college mental health services, which reveals how long-standing medical discourses concerning student wellness work to individualize what are often structural and systemic factors of ill health. It then provides an account of the ways the application is part of public universities' participation in what I have called "big data socialization." In the case of WellTrack, the app socializes students to constantly engage, interact, and relinquish data in exchange for health services while simultaneously promoting a depoliticized understanding of mental health within the university.

The WellTrack app typifies how notions of choice, agency, and empowerment are used to associate self-surveillance with a form of self-care that neither acknowledges the structural forces contributing to student mental health issues nor is oriented toward structural change. Structural inequality impacts and conditions student wellness, and yet these forces are obfuscated by the WellTrack app's design, which emphasizes individual behavior abstracted from social structures of power. As communications scholar Sasha Costanza-Chock explains, intersecting inequalities can shape all stages of the design process, "including (but not limited to): designers, intended users, values, affordances and disaffordances, scoping and framing, privileged design sites, governance, ownership, and control of designed objects, platforms and systems, and narratives about how design processes work."[4] The following describes the ways that intersecting inequalities are reproduced through the WellTrack app's affordances and disaffordances, presuppositions about its users, embedded values, and data collection practices. I emphasize how the experiences and circumstances of marginalized students are largely rendered outside the frame of WellTrack's design considerations.

This chapter helps us identify the underlying assumptions and contradictions in the design, marketing, and media reception of the WellTrack app's introduction into North American universities. It provides an account of the ways that public universities are rationalizing the proliferation of digital technologies across multiple spheres of life. The focus of this chapter's analysis is to map the ideological values embedded in the promotional discourses and design of the WellTrack app alongside the concrete social, historical, political, and economic forces underpinning its development and implementation in higher education. Ultimately, this chapter shows how the app's design and the promotional discourses that shape its reception encourage students to engage in constant self-examination and to strive toward a vision of student wellness that precludes an analysis of the structural conditions contributing to poor student mental health, including austerity and racism.

Brief History of College Health Programs

The first college health program was founded at Amherst College in 1861. This program focused primarily on students with physical illnesses, and physical exercise was recommended as a way to circumvent emotional distress. Faculty or clergy, not health center staff, would provide counseling, and it was often the case that emotional problems were, as psychiatrist David P. Kraft explains, "judged to be spiritual or moral deficiencies beyond the purview of medical specialists."[5] With the rise of the mental hygiene movement in the early twentieth century, clinical psychology and social work came to influence conceptions of appropriate mental health services for students. In contrast to earlier conceptions that treated what is now called poor mental health as a spiritual or moral deficiency, the mental hygiene movement focused on individual maladjustment to environmental factors. This discursive shift produced the idea that students suffered from "personality development" problems, which were resulting in well-qualified applicants leaving school and failing to complete their studies. Princeton University went on to create the first dedicated mental health service for students with personality development problems in 1910. From the inception of mental health services for students, retention was a primary concern and framing device for understanding how students could be harmed and healed.[6]

This focus on retention is typified by American psychiatrist Frankwood Williams's 1921 address at the first meeting of the American Student Health Association. Williams stressed the importance of student retention; the minimization of student "inefficiency," "mediocrity," and "unhappiness"; and the need for social control as reasons for establishing mental health programs in colleges.[7] These points of emphasis speak to the ways that sociomedical discourses produced ideas of the healthy student, which corresponded to an institutional interest in retention as well as dominant ideas of workplace discipline (efficiency and self-control) in order to generate docility and compliance with university expectations. Additionally, for Williams, political radicals on college campuses were "university causalities" whose "mental integrity" was open to question:

> Very many of these young radicals ... [are] students whose intellects and whose physical condition have been carefully attended to, but whose emotional lives and habits have been permitted to take their course. Finding no other suitable outlet, emotional energies generated at sources quite apart from and bearing but slight if any relationship to the situation at hand (usually quite ascertainable sources) have flown into these social situations.[8]

For Williams, student mental health services would support the college socialization process, helping students to discipline their emotional impulses. Williams created a clear separation between the individual, whose emotional impulses needed to be treated and managed, from the social situations in which these impulses were being expressed. Student activism would go on to surge in the 1920s in response to gender- and race-based segregation on college campuses, mandatory daily chapel, and room inspections, particularly at historically Black institutions.[9]

For Williams, the university's job was to train students' emotions as well as intellect, and thus physical and mental health had to be part of the university's program.[10] However, Williams was careful to distinguish between those who suffered from "feeblemindedness" and "insanity" and those the mental hygiene movement would serve within the university. In providing an account of why universities had been indifferent to the mental health of their students, he stated:

> Mental ill health has meant to them mental deficiency (feeblemindedness) or mental disease (insanity), and of the former there is none to be found in universities and of the latter only an occasional case. Mental hygiene, therefore, cannot be an important

problem for them. This, however, is a misconception. Mental hygiene as a movement has had to concern itself very largely with the problem of the care and treatment of the great body of helpless sufferers ill of frank mental disease and with the social and economic problems that develop about them; but mental hygiene as a department of medicine is vastly more concerned with the mental health, the happiness, and the efficiency of the average normal person, of you and of me, of our wives and our children and our neighbors.[11]

Although the mental hygiene movement differed from eugenics in highlighting the ways that social and environmental factors could impact one's mental health, Williams was also working to avoid alienating university administrators and his peers in the medical field by distinguishing between "helpless sufferers ill of frank mental disease" and the "average normal person." Eugenics was a widely accepted theory of biological determinism in the 1920s, which is why many universities were apprehensive to start a mental hygiene program, wanting to preserve their status as institutions for training society's elite.

The early college mental hygiene movement argued for an understanding of mental illness within universities as a direct result of students' poor adaption to the expectations and demands of the university setting. This is due, in part, to the influence of Meyerian psychobiology on the collegiate mental hygiene movement. Adolf Meyer, and the mental hygiene movement's founder, Clifford Beers, were key partners in ushering in mental health reform in the twentieth century. However, even while the movement acknowledged that university life created new challenges and stressors, it was the individual student's inability to adapt that the mental hygiene movement sought to address. As historian Sol Cohen explains,

"The mental hygiene movement was basically conservative in its emphasis on the individual and in its turning away from social or political change . . . There was no need for major social overhaul or institutional change. The problems facing the nation were those of individual personality."[12] More than a century after the publication of Beer's memoir *A Mind That Found Itself* in 1908, which outlined the mental hygiene movement's approach mental health, university ideas of wellness persist in socializing students to think of mental health as a matter of individual, personal responsibility.

Student Wellness under WellTrack

The remainder of this chapter investigates the ways that the WellTrack app, and more briefly, Amazon Echo Dots in student dormitories, participate in the construction of what constitutes student wellness, while simultaneously operating as a form of big data socialization. Since the collegial mental hygiene movement, student wellness initiatives have emphasized encouraging students to adapt to the university's demands for efficiency and wellness. With WellTrack, the app encourages students to continually self-manage their moods, thoughts, and behaviors in a digital environment designed for data capture. Students' self-understanding of their own wellness is produced, in part, through their interaction with the application and its metrics. Metrics, as sociologist David Beer explains, "shape the ways that the social world is understood, approached, and how the people that constitute it are classified and categorised."[13] These metrics inform how students understand desirable health outcomes in the context of rising rates of mental health struggles in higher education.

The WellTrack app is a technology of subjectivation, meaning "a complex set of processes whereby one's sense of self as an individual agent is paradoxically shaped according to processes and forces external to the self. Such factors include not only race, class, and gender, of course, but also social norms, educational circumstances, labor potential, and health."[14] As this chapter demonstrates, dominant institutional imperatives under neoliberalism shape how the WellTrack app produces ideas and practices of self-knowledge about student mental health, which discount structural factors of ill health, including social and economic inequality. Furthermore, as argued above, the promotion of individualized practices of self-care is also deeply rooted in long-standing conceptions of student wellness in US universities. It is through ideas of how to care for oneself that individuals come to internalize norms for what is deemed appropriate conduct. To be a "good" student is to pursue self-knowledge with the aim of being responsible and self-regulated.[15]

The WellTrack application must also be understood as part of the larger movement of self-tracking. From private businesses to universities, self-tracking tools are being used as a means of preventative healthcare. This emphasis on individual self-management is part of the political rationality of neoliberalism, which centers self-responsibility and the market economy as a means of rationalizing the withdrawal of state support for social programs. Rather than relying purely on top-down mechanisms of social control, neoliberal political rationalities also incentivize the voluntary participation of subjects in the discipline and management of their own thoughts and behaviors.[16]

This emphasis on voluntary individual self-management is apparent in university press releases and news reports regarding the adoption of the WellTrack application. For instance, Emily Mertz from *Global News* describes how the "Canadian-made wellness platform is helping students at the University of Alberta become more aware of and engaged in their own mental health."[17] Students are responsible for monitoring themselves, and this act of self-monitoring is conceptualized as responsible self-governance. Olivia Ginsberg's report in the *Miami Hurricane* presents WellTrack as a convenient solution for improving mental health for students who cannot or will not go to the counseling center.[18]

Implicit in many proponents' celebratory accounts is that the rising costs of healthcare, increasing demands for mental health resources, and resource limitations are impossible barriers to transcend, so technologies like WellTrack can provide an immediate workaround. One such example is Jodie Vanderslot's write-up in York University's community newspaper, *Excalibur*. Here, the WellTrack application is presented as a solution to rising healthcare costs and demand for mental healthcare. According to Vanderslot, the app reaches students "in a way that may not otherwise be possible."[19] Similarly, Gaya Arasaratnam, acting director of strategic projects for student health and wellness at Ryerson University asserts, "if a student needs help at 10 p.m., they can get it instantly with WellTrack."[20] Students who do not have the resources or time, who feel stigmatized in seeking out help, or whose universities have extended waiting lists can instead turn to an application for assistance. These are barriers that marginalized students, and particularly low-income students, students of color, queer, and gender nonconforming students, disproportionately face because of existing inequalities in health services and outcomes, which are compounded by the financial limits of campus wellness resources.[21]

Although a readily available mental health app may help mitigate access barriers in the short term, given that it is offered to university students at no additional cost and available on demand, it is important to critically attend to the *type* of therapeutic care that the application offers. This includes evaluating the relationship between therapeutic care and desired subjectivating practices under neoliberalism, as well as whether the therapeutic techniques on offer account for, and address, structural factors impacting students' mental health.

Cognitive Behavioral Therapy and Neoliberalism

While cognitive behavioral therapy, the therapy on which the app is modeled, has been shown in some studies to be an effective treatment for symptoms of depression and anxiety, the overall efficacy of CBT is greatly contested.[22] Furthermore, CBT delivered through low-cost, online measures involving self-administration and mindfulness techniques has been shown to have "high dropout rates" and "no long-term benefits."[23] Nonetheless, it is increasingly advocated for in a range of institutional contexts, including schools, workplaces, and prisons. Even prior to CBT's digital proliferation, its design as a therapeutic technique has been consistently compatible with the political imperatives of neoliberalism. Psychiatrist Aaron T. Beck's development of CBT in the 1970s coincided with new demands from the US Congress and insurance companies for "quantitative proof of therapy's efficacy and cost-effectiveness," in tandem with rising healthcare costs and the increasing power of private health insurance companies in the United States.[24] This therapeutic framework was offered in response to political and economic pressures for psychotherapy to adopt hyperrational forms of assessment, particularly randomized control trials and actuarial modeling.

CBT has also been critiqued within contemporary clinical psychology debates as a highly individualized form of treatment. For instance, historian Åsa Jansson argues that CBT is one of the favored treatments for affective disorders under neoliberalism because "an individualised, neurobiological model of psychological distress sits comfortably within a political framework that emphasises individual responsibility and choice over social support."[25] Similarly, healthcare ethics researcher David Ferraro argues that CBT is grounded in an individualist, disciplinary approach that seeks to "inculcate correct thinking, behaviour, and self-observation... applicable to anybody and amenable to administration by a computer programme."[26] The decision to build the app around CBT could also be informed, in part, by a desire to generate the information that would allow app developers to best assess users' engagement with the application itself. It is engagement, after all, that allows software companies to demonstrate the efficacy of their technologies to potential investors, advertisers, and consumers.

This compatibility between the therapeutic protocols of CBT and the political framework of neoliberalism is evident in the WellTrack's self-described key function, the Mood-Check. As Deborah Lupton explains, neoliberal political rationalities "generally rely on apparatuses of 'soft' rather than 'hard' power. Instead of relying on coercive measures that appear to be imposed from above... neoliberal political systems invest faith in the voluntary take-up of imperatives by the citizens themselves."[27] WellTrack sends users daily nudges to log their mood and describe what is on their mind in order to receive, according to the app, "the right care at the right time." To document their mood, the user drags a circular icon until it lands on the emoji they think best captures their mood. There are also options to select corresponding words for the mood, such as "wonderful," "excited," "happy," "stressed," "tired," "overwhelmed," and "I don't know." The user then indicates what they are currently doing, whom they are with, and their location. This information is used to produce a data point on a Mood Heatmap, which shows the user how their current mood compares to past logged moods.

This form of mapping, like other self-tracking techniques, allows the user to make sense of their mental health through an externalized visual representation. Thus, the mood mapping function, and the promise of greater data-driven insights into the self, persuades users to engage in self-management and to routinely invest time, attention, and information into the application. Additionally, as feminist human-computer interaction scholars have pointed out, self-tracking designers and engineers frequently rely on reductionist categories, such as "mood scores," because they are easily trackable through data.[28] WellTrack relies on a discrete, trackable unit—a mood in a given moment in time—that is abstracted from an account of the structural forces contributing to student mental health outcomes.

If we consider the WellTrack app not in isolation but as a sociohistorically and politically situated tool, we must attend to the ways it is tangled up with broader shifts in the structure of the university, work, and capitalism. For instance, WellTrack's emphasis on digital self-care and self-monitoring mirrors the marketing of self-tracking surveillance technologies in emerging corporate wellness and interactive life insurance programs, which produce data that can be sold for refining targeted advertisements, products, and services. These types of technologies also empower companies to enact punitive consequences for users who do not or cannot exercise, eat well, or sleep well.[29] While students are being encouraged to self-track for enhanced mental wellness and expected to demonstrate affects of resilience, flexibility, and grit under conditions of rising social and economic precarity, digital management systems are demanding that workers self-track their behaviors, attitudes, moods, stress levels, and practices for the purposes of increasing their agility and flexibility. In the workplace, these digitization methods have been shown to lead to "high turnover rates, workplace rationalisation and worker stress and anxiety," which expert on worker rights and technology Phoebe V. Moore links to rising levels of both objective and subjective precarity.[30] In both cases, the affective capacities of students and workers are being channeled toward developing the desires and habits that neoliberal capitalism requires.

In addition to MoodCheck, users are also encouraged to use the Thought Diary tool to analyze events that cause them to feel anxious or depressed. Users are asked to identify a specific event, situation, or thought that produced anxiety or depression, describe the event, list their feelings, and rate the intensity of these feelings on a scale from 0 to 10. Users are then asked to identify the thought that produced these feelings and to select from a list of unhelpful thinking styles that correspond to the thought. Afterward, users are asked to challenge these thoughts by questioning the validity of their emotional experiences using prompts such as, "What is the likelihood of that happening?" and "If I wasn't depressed, how might I view the situation differently?" Much like the twentieth-century discourse of the mental hygiene movement, students are encouraged to manage their emotional responses to phenomena. However, with WellTrack, these emotional responses can now be quantified and analyzed en masse to produce overall assessments of campus wellness for counseling center directors. From the early collegial mental hygiene movement to the WellTrack app, it is the possibility of collectively identifying inadequacies and flaws in the university itself that are consistently obscured.

WellTrack, Structural Racism, and Austerity

WellTrack's emphasis on individual self-care and accountability is just one way that the app serves to disconnect student mental health from structural factors. In the application's Cognitive Distortions Quiz, the express purpose of which is to help students understand the different thinking styles that WellTrack deems "unhelpful," it becomes clear that WellTrack has embedded presuppositions about its users that fail to account for the ways that economic and racial inequalities inform students' life experiences. For instance, the second quiz question reads as follows:

> Natalie is leaving a store without having purchased anything. As she is about to pass through the anti-theft devices at the door, she thinks, *"What if the alarm sounds, even though I haven't taken anything? What if the security comes and they accuse me of stealing? What if I'm treated like I did something wrong?"* What kind of distortion is Natalie using?

According to the app, Natalie is engaging in "catastrophizing— she is imagining an unlikely and extreme scenario, which is then causing her to feel anxious." The app then presents an example of an undistorted thought in the given scenario: *"I havent'* [sic] *taken anything from the store, so the alarm won't sound."* What this quiz question reveals is that, either the unmarked student the application has in mind is white, or the designers of the application's quiz function are not aware of the routine forms of discrimination, intensified conditions of surveillance, and higher rates of police violence that people of color face. Scholars of race and technology Simone Browne, Dorothy E. Roberts, and Ruha Benjamin have stressed that surveillance is a key technology of racial oppression, from the transatlantic slave trade to contemporary mass incarceration.[31] People of color are frequently racially profiled and criminalized during routine, everyday encounters with surveillance workers such as security guards, and African Americans file the majority of false arrest complaints in US retail settings.[32] The app reinforces the message that what might be a legitimate response to structural conditions of racism and inequality is in fact a cognitive distortion.

This example of Natalie is a symptom of an underlying problem with WellTrack's design, as well as with broader digital education governance practices that understand students as separate from their corporeality as raced, classed, and gendered subjects, abstracted from the social contexts that inform their lived reality. The WellTrack app provides no explicit guidance for students experiencing structural and interpersonal racism on campus. For example, mismatch theory, grounded in racist assumptions about aptitude, argues that students of color are harmed by policies such as affirmative action because they are then matched with schools that are too competitive for them. Some mental health experts have suggested that Black students "who strive to simultaneously excel in the classroom and disprove the mismatch theory might ultimately overwork themselves to the point of illness" to demonstrate intellectual worth.[33] Universities across the United States, which are rooted in histories of racial-colonial oppression, are persistent sites of discrimination, xenophobia, tokenism, and surveillance for students of color. Living in a society constituted by white dominance has corporeal and mental effects that go unaccounted for in dominant race-neutral discourses concerning individual students' "grit" and "resiliency."[34] As education scholars Ebony O. McGee and David Stovall explain, "Only the eradication of racism will alleviate race-related stress for African Americans and other historically racialized populations."[35] And yet psychic phenomena within the WellTrack app are matters of individual thoughts and behaviors that can be reduced to a set of cognitive distortions and unhelpful thinking patterns, rather than responses to the pervasiveness of racial and class oppression in higher education.

While the WellTrack app's design stifles opportunities for students to think critically about university conditions, it operates within an institutional context shaped by both structural racism and austerity. Americans currently owe over $1.75 trillion in student loan debt, which is roughly $719 billion more than the total US credit card debt.[36] Additionally, Black students borrow federal student loans at higher interest rates, borrow more at public colleges than other students, are more likely to need to take on debt to earn a college degree, and are more likely to have difficulty repaying student loans.[37] These statistics reflect structurally racist employment, wage, and educational access disparities, which lead to families of color having fewer resources to pay exorbitantly high tuition rates at public universities. Black students have less generational wealth due to histories of slavery, redlining, and labor inequalities; are more likely to be the first in their family to go to college; make less money after graduation; and are more likely to drop out.[38] For many students of color, the impacts of racism on student mental health discussed above are compounded by experiences of financial precarity.

There is one instance where debt is explicitly addressed in the app, and this is through the Activity Scheduler tool. This tool allows students to schedule activities that are categorized based on whether they are "Pleasurable," "Social," or "Achievement" oriented. For achievement activities, the list of provided examples includes doing household repairs, exercising, cleaning, returning library books, writing a thank-you note, and paying debts. The app encourages students to have a positive response to the act of paying into debt, which hardly challenges the increasing rollback of public funds to support higher education or makes room for collective struggle against financialization.[39] The framing of paying off debt as an achievement, which rewards students with praise, promotes student acquiescence to the increasing rollback of state and federal funds to support public education.

These forms of individualizing self-care that WellTrack encourages students to adopt stand in stark contrast to the self-care practices envisioned by Black feminist scholar Audre Lorde in the epilogue to *A Burst of Light*. Lorde situates self-care as a strategy of resistance against interlocking forces of oppression that shape the lives of marginalized people: "Caring for myself is not self-indulgence, it is self-preservation, and that is an act of political warfare."[40] In Lorde's formulation, self-care cultivates a combative form of resilience in the face of social and political forces that seek to stifle one's ability to survive. It is important to heed feminist theorist Sara Ahmed's warning about treating all forms of self-care as inherently neoliberal, which collapses the distinction between self-care practices indebted to the tradition of Black feminism with the forms of self-care that perpetuate dominant socioeconomic paradigms.[41] WellTrack, however, promotes an understanding of self-care that is separated from an account of the social and economic forces contributing to mental illness, which disproportionately harm marginalized students. Rising levels of precarity under neoliberalism are what intensify insecurity, subsequently increasing efforts to anticipate and predict.[42]

Data Capture and Consent

WellTrack's design for self-tracking, like those of most digitized commercial self-tracking apps, doubles as a form of capitalist dataveillance, meaning surveillance that uses technology to generate digital data that can be captured, monitored, and exploited for profit seeking.[43] While users of this app engage in self-care in that they record information about themselves to optimize and improve their mental health, the app also monitors and collects information about users for commercial gain. WellTrack's deployment of dataveillance raises issues concerning student privacy and data exploitation that dominant notions of consent, enshrined in privacy policies, do not address.

Students who want to try out the application services described above must first agree to WellTrack's privacy policy. In general, the term "privacy policy" is misleading. These contracts between users and companies do not guarantee that the user's data will be kept private but instead stipulate what information a given business gathers and how it collects, uses, or shares that information with others. WellTrack's privacy policy stipulates, "We will not actively collect Personal Information for the purposes of sale or marketing in a way that specifically identifies the individual. We do not sell consumer lists."[44] The use of the word "actively" leaves open the possibility that there might be passive forms of personal information collection. Additionally, the qualification that the company does not collect personal information that specifically identifies the individual suggests that it may collect personal information that is then aggregated and anonymized. Aggregate and anonymized data can still have very real effects on individuals, and these effects often reinforce existing inequalities and asymmetries of power.[45] One such example is the case of Crisis Text Line, a nonprofit, AI-assisted chat service for suicide prevention and crisis intervention, which monetized anonymized text data generated by people experiencing immense psychological and emotional distress through a data-sharing agreement with Loris, a customer service company.[46]

Additionally, if student data is being collected, anonymized or not, for the purposes of sales or marketing, then students are directly involved in the company's profit-generating mechanisms by virtue of using a mental health service their university has offered. It's also the case that student data is put to use for what previously might have required a company to invest in market survey research, such as which aspects of the app are being used and how often. Student data, then, is a means by which companies can potentially reduce costs, while simultaneously serving as evidence to support market expansion. This is part of a broader pattern in technology for the higher education sector, where increasingly, corporations, researchers, and universities use student data in extractive ways with varying levels of student consent (and in some cases, no informed consent at all).[47]

WellTrack's privacy policies state that consent may be withdrawn at any time. However, the withdrawal of this consent also means that a user would no longer be able to continue using these services. For students who might rely on WellTrack for mental healthcare, regardless of the app's effectiveness or the drawbacks discussed here, this bargain can take on a coercive character. Those who are most vulnerable and unable to access alternative care are in less of a position to forgo using the service. Additionally, WellTrack introduces users to third-party vendors and websites, which may deposit cookies on the user's computer, the privacy policies of which are not under their control. Even for users who decide to terminate their relationship with the company, this does not guarantee their personal information will be erased. In fact, by agreeing to WellTrack's policies, "you acknowledge and agree that if you request that your name be removed from our databases, it may not be possible to completely delete all your Personal Information due to technological and legal constraints."[48] WellTrack is symptomatic of ways that the larger tech industry unethically shares private customer health information with third parties through contracts with users.[49] If student data is being collected, anonymized or not, for the purposes of sale or marketing, then the domain of students' thoughts, moods, and behaviors is involved in private companies' profit-making mechanisms by virtue of using a university mental health service.

WellTrack's privacy policies also state that in certain cases personal information may be disclosed without user consent if "the disclosure is to a person who requires the Personal Information to carry out an audit for, or to provide legal services, error management services or risk management services to CyberPsyc" or if "CyberPsyc reasonably believes that the disclosure is required to prevent or reduce a risk of serious harm to the mental or physical health or safety of the individual to whom the information relates or another individual."[50] There is no description of what counts as "serious harm" to mental or physical health, and therefore, CyberPsyc is able to exercise discretionary power over if and when it will disclose the user's personal information, depending on the circumstances.

Certainly, there are aspects of WellTrack that provide useful information for students and tools they might find helpful. However, it is important to situate these tools within the larger sociopolitical context of the application's use, as well as the app's relationship to dataveillance. For example, there is a Zen Room that facilitates meditation sessions, complete with a range of ambient noises to choose from. However, as sociologist Rosalind Gill and social psychologist Ngaire Donaghue note, it is ironic that "a set of techniques and practices developed from a tradition that is critical of Western achievement orientation, [is] now being enthusiastically embraced in an effort to soothe some of the harsh psychic consequences of the always-on, constantly striving, contemporary academic culture."[51] Furthermore, when students fill out the app's wellness exam, particular questions will provide students with campus-specific resources, depending on whether they click yes to questions such as, "are you experiencing academic difficulties?" and "are you worried about your financial situation and/or have concerns about your basic needs (housing/food/transportation)?" However, based on the student's wellness exam, they are referred to in-app CBT modules, which result in further data collection. This information about the resources that correspond to the user's potential needs is part of a trade-off for information that the private firm collects.

Part of the way WellTrack markets itself is by emphasizing its ability to offer campus-specific resources to students. The WellTrack website explains that universities can integrate their "on- and off-campus resources for academic, financial, health and wellness, substance abuse, safety and security, and sexual violence into WellTrack. Individual students will be recommended a customized suggestion based on which areas they indicate they need more support."[52] The results students see are tailored to the services their institution has to offer and the aspects of mental health they indicate needing support for. This promise of relevancy and customization is similar to how other firms, including Amazon, have successfully established contracts with universities despite privacy concerns.

WellTrack and Amazon Echo Dots

In 2018, Saint Louis University became the first university in the United States to bring Amazon Alexa-enabled devices into every student residence hall room and student apartment on campus. This contract included the deployment of thousands of Echo Dot smart devices, customized to answer more than a hundred questions specific to Saint Louis University. According to the campus's chief information officer David Hakanson, "the new tool can boost efficiency," helping students to get responses immediately rather than having to search the web.[53] Here, ideas of both relevancy and increased efficiency helped shape the technology's integration into campus life.

In a university culture that places increasing expectations on students' time and attention, the integration of this technology is perhaps unsurprising. However, it's important to acknowledge how universities enact discretionary power through these tools when it comes to determining what questions about the university are answerable. For instance, a student reportedly asked, "Hey Alexa, why is tuition so high?" Alexa replied, "Hm, I don't know that one."[54] Other universities to begin including Echo Dots in dormitories include Arizona State University, the Georgia Institute of Technology, and Truman State University. Northeastern University has partnered with a start-up company called N-Powered to develop an Alexa "skill," the industry term for apps that give Alexa more capabilities, which will allow Echo Dots to answer questions specific to a given student's schedule and other personal information. To get this skill set up, students need to release their rights to their information using a Family Educational Rights and Privacy Act waiver. On some campuses such as the University of Texas at Dallas, students have raised concerns about their lack of inclusion in decision-making processes regarding these tools and expressed discontent through campus newsletter polls about university plans to install Echo Dots in dormitories.[55]

Not only are the WellTrack app and the Echo Dot both marketed to universities using notions of relevancy, but also, both of these digital services are presented as having a relationship to student wellness. For example, in the promotional video for Saint Louis University's decision to put Alexa-enabled Echo Dots in student dormitories, "Welcome to SLU, Alexa," a student is depicted asking questions related to campus events, the library and student health center hours, food delivery services, guided meditation, help with studying, career services, and, ultimately, the time of commencement. The promotional video presents Alexa, and the university-branded Echo Dot from which the Alexa voice comes out, as an essential part of the student's journey to success and a source of support for the student's well-being.[56] The video also advertises Amazon's paid subscription service, Prime, as the student is depicted opening up an Amazon Prime package to find a pair of winter gloves. The student then asks Alexa to call their mom, presumably so that they can thank her for the thoughtful gift. In this way, the promotional video frames the Echo Dot as a tool for supporting relationships of love and care between students and their parents.

In another promotional video for the Echo Dot, in this case for N-Powered's "student helper" skill, the camera focuses on a university-branded Echo Dot, with a student's voice from outside the frame asking questions.[57] When "Ryan," the featured student in the video, asks Alexa to "start student helper," the school's anthem begins to play. Alexa goes on to answer Ryan's questions about laundry, their financial balance, and ultimately, who Ryan "is," to which the device replies, "you, my friend, are Ryan Lucarthy. Should you ever forget your name again, I am here to help you find your way." The student helper skill is designed to nudge students' behaviors in ways that are consistent with the university's linking of student wellness to academic success. Such nudging encourages students to self-govern in ways that are consistent with university-oriented norms of self-care.

This partnership between Amazon and universities raises two key ethical problems—that of exploitation and that of privacy. By collecting voice inputs to improve and develop the quality of Amazon's artificially intelligent services, students' voice data becomes a form of unremunerated crowd-sourced labor for product improvement. This renders students' data highly profitable for Amazon, given that they are not directly compensated for their assistance in the product development process. Additionally, at ASU, there are initiatives to help foster further connections between students and Amazon's brand identity. For instance, ASU created a new $5,000 scholarship to award winners in an Echo Dot Hackathon, which came with a tour of Amazon's Seattle headquarters. There have also been reports that Amazon is considering introducing targeted advertising into their Alexa services, which could make universities an instrumental part in delivering student audiences to the company (although Amazon has denied it has plans to introduce its own advertising). However, Alexa currently allows companies such as streaming and radio services to have ads, so long as they do not imitate the voice of, or refer to, Alexa.[58]

Amazon's Echo Dot also raises many privacy concerns. Once students use the "wake word," "Alexa," Amazon keeps a recording of what users say and Alexa's reply. While it is possible to delete the record of these commands, this information is not generally part of how universities explain the device's function and purpose, nor is it emphasized in the Echo Dot's marketing. There have also been reports of Alexa recording a conversation without the wake word being given and sending the conversation recording to another user.[59] Finally, it still remains to be seen whether the US government will be able to access Alexa recordings without a search warrant, but legal arguments have been made that these recordings fall under the "third party doctrine," which stipulates that the Fourth Amendment does not protect personal information shared voluntarily with others, such as a bank or internet provider. Although the 2018 decision in *Carpenter v. United States* might establish a precedent for data collected by smart dorm technologies, having ruled that the government's warrantless acquisition of cell-site records violated the Fourth Amendment, this is not yet constitutionally guaranteed. Given pervasive police repression at campus strikes and student protests, recordings could render students involved in organizing efforts more vulnerable. Currently, there is an Amazon webpage where law enforcement can fill out a form, state there's a life-threatening emergency, and get access to user data without consent or a court order.[60]

While universities give students the option of unplugging and storing the devices, this reduces concerns over privacy and exploitation to a manner of individual choice. The default assumption of the universities providing these devices tends toward placing them in dormitories, rather than having students request them. These partnerships between private firms and universities lend themselves to big data socialization, where interfacing with these technologies is promoted uncritically, and where the ideological value of efficiency is inculcated within students. Hakanson describes how Alexa will help reduce inefficiencies, such that students can get responses to their campus-related questions "in five seconds" rather than "two to four minutes."[61] Thus, the presence of these devices on campuses reinforces the notion that students need to habituate themselves to constant expectations on their time and attention. Jonathan Crary describes this as part of the structure of 24/7 capitalism, in which there are no minutes to spare. It is an injunction to always be active, and therefore interactive, where "one's life is coinciding with whatever applications, devices, or networks are, at any given moment, available and heavily promoted ... Committing to activities where time spent cannot be leveraged through an interface and its links is now something to be avoided or done sparingly."[62] Both the marketing of Echo Dots and WellTrack as improving inefficiencies for students reinforces the idea that time should be spent, and can be maximized, with the use of digital technologies.

The Future of Digitized Student Wellness

Despite the concerns identified above, campus health centers continue to contract with CyberPsyc Software Solutions. One reason for this is that WellTrack provides data analytics that tell health center directors which issues, according to the app, are "most prevalent on campus, and how many students have improved their mental health using our self-help programs. Insights also summarize resource use."[63] Given that funding is the number-one challenge mental health initiatives face on college campuses, it is not surprising that public universities are turning to data analytics to help manage resources. Additionally, as *USA Today* reporter Caroline Simon explains, "a desire to help students isn't the only motivating factor for addressing mental health. Healthier students lead to higher retention rates and graduation rates."[64] Institutional metrics significantly influence how universities approach student mental health considerations.

The WellTrack app is also marketed essentially as a cost-savings measure, encouraging students to engage in the forms of mental health self-management that the app promotes rather than increasing resource-intensive on-campus counseling services. Steve Jenkins, sales development representative for WellTrack, explicitly positions the app as a solution to the problem of the "overutilization" of campus mental health resources in sales communications.[65] Furthermore, he writes that "universities are typically early adopters of technology so there is a chance that the successes that these universities have may trickle over into the corporate world. Especially as students become tomorrows [*sic*] employees."[66] Jenkins links, quite explicitly, enhanced student wellness with enhanced corporate gains. This framing is in keeping with larger international trends in higher education that seek to equip young people with emotional management skills so that they become more "accountable" for their role in the labor market under rising conditions of economic precarity.[67] It is not clear how WellTrack helps universities deliver what students report will be most beneficial to their mental health—the reduction of stigma, which requires collective action to redefine toxic cultural norms and improved faculty and staff education about mental health issues.[68] Although university contracts with CyberPsyc Software Solutions do not preclude these other forms of outreach, as demonstrated above, the notion of wellness the app facilitates makes students individually responsible for constant self-management and self-monitoring in a software environment designed for data capture.

This focus on individual conduct produces a limited frame of what seems possible to transform about higher education, sidestepping students' capacity to challenge the underlying socioeconomic, political, and institutional structures that impact mental health. If wellness app developers were to incorporate design justice principles, perhaps this could lead to the development of nonexploitative practices that limit the power of the app designer and instead empower students—those directly impacted by the design process—to direct its goals and outcomes.[69] In the context of university wellness apps, this approach might include marginalized students directly guiding the design process and ensuring that people committed to racial and economic equality have control over the distribution of its benefits. However, as Ruha Benjamin warns us, the rubric of design "could also sanitize and make palatable deep-seated injustices," including but not limited to the racial, economic, and gender injustice underpinning so much student suffering.[70]

WellTrack was featured on several universities' counseling center resource webpages for coping with stress and anxiety after the first wave of COVID-19. The exponential growth in the use of digital tools during the pandemic will likely continue to increase the diffusion of data analytics and private-public partnerships for digitally monitoring student behavior. Furthermore, the pandemic has compounded existing conditions of financial instability for many universities caused by decades of austerity and privatization. These institutions are likely to find promises of technology-fueled cost savings and efficiency particularly attractive in the years to come.[71]

If universities are to be places that support students' mental health, technological solutionism needs to be contested. Instead, efforts must push for the radical social and economic reforms necessary for meaningfully addressing student mental health in a socially unjust society. This includes collectively resisting universities' complicity in neoliberal policies and long-standing investments in individualist understandings of student wellness, with disparate consequences for those living and studying at the university's margins. The next chapter shows how campus security gets constructed historically and presently such that the university is seen as what needs securing through the production and application of digital technologies, rather than as a source of students' financial insecurity and vulnerability to state repression.

Chapter Five

Security

On September 3, 2020, the exam proctoring software company Proctorio served a lawsuit to learning technology specialist Ian Linkletter after he made a series of critical tweets about its software, which included several links to unlisted YouTube videos from Proctorio's help website for instructors. According to Linkletter, he did so out of concern for the ways this software requires students to subject themselves to artificial intelligence that analyzes their physical bodies and any audio coming from the room while taking a given test.[1] During an exam, Proctorio works by recording a given student's computer camera, audio, and the websites they visit; measuring a student's body language and behaviors; and flagging any movements or actions that the software considers suspicious using machine learning and facial detection technology. These flagged moments are then given a color-coded Suspicion Score, meaning a rating of the degree to which academic misconduct is suspected, which professors use in tandem with watching the recording of a given student to decide whether cheating has occurred.[2]

A range of faculty, students, and education technology experts have argued that Proctorio is essentially spyware. It can reportedly read and change the data on students' web browsers, modify keyboard functions, store keystroke movements, capture all screen content on a student's computer, identify all devices connected to students' computers, change all privacy-related settings on students' computers, monitor eye movements via webcam and save recordings, record and store all sounds while in use, and require initial and periodic room scans.[3] Proctorio is just one of several well-known companies that make up the digital proctoring industry, such as ProctorU, Examity, and Verificient. These tools can require students to show a form of photo identification that includes a variety of sensitive personal data, such as citizenship status, which is then matched to their "biometric faceprint" as captured by their laptop camera. Exam proctoring software hinges on baked-in assumptions about what compliant test-taking behavior looks like, as well as the notion that identity markers are fixed and immutable for biometric student identification.

Both students and faculty have argued that exam proctoring software disparately impacts marginalized students, including arbitrary flagging students with darker skin tones, physical disabilities, and internet connectivity difficulties, as well as students with children, neurodivergent students, trans students, and students with religious head coverings.[4] For marginalized students, their appearances, behaviors, and testing environments are more likely to fall outside of the exam proctoring software's models for "normal" behavior and appearance markers. This is due to the algorithmic privileging of able-bodied, white, cisgender students with access to private spaces and reliable internet connections. Despite these concerns and critiques, Proctorio experienced a 900 percent increase in business during the first few months of the COVID-19 pandemic.[5]

Since the beginning of September 2020 when the lawsuit against Linkletter was first filed, hundreds of university faculty, staff, administrators, and students from a range of countries have signed an open letter in his defense. Miami University student Erik Johnson also took to Twitter to criticize Proctorio for privacy-invasive tactics, a lack of transparency, and practices that discriminate against low-income families through an analysis of Proctorio's source code. In response, Proctorio filed a copyright takedown notice.[6] Student-led movements at campuses that include Linkletter's home institution of the University of British Columbia, as well as US public universities such as Miami University, the University of Colorado Boulder, the University of Illinois Urbana-Champaign, and the University of Minnesota, have organized against invasive exam proctoring software on their respective campuses, either through demands for limits or bans on the technology. I look at these movements more closely in the final chapter.

While education technology companies' litigiousness and the discriminatory and privacy infringing dimensions of exam proctoring software are alarming, this chapter focuses on the question of what it would mean to read Proctorio, and similar tools that use machine learning and biometrics to detect "cheating," as symptoms of a larger, diffuse logic of securitization that underpins the smart university. Securitization is "a set of interrelated practices, and the process of their production, diffusion, and reception/translation that brings threats into being."[7] The ideology of securitization that this chapter sets out to map is grounded in the following principles: that the university's continued viability as an institution is contingent on preserving the security of the grading system, that the security of the university hinges on preventing attacks from within and without, and that the university has an institutional obligation to participate in large-scale domestic and global security efforts through technological innovation. Through securitization, historically and presently, the university normalizes and legitimizes US racial and economic injustice, including through the enforcement of austerity measures. Contemporary digital technologies marketed and deployed in the name of improved university security, such as smart lighting, college safety apps, video analytics technologies, and exam proctoring software, must be understood within longer histories of why and how universities enact security within and outside their borders of belonging.

Securing the Grading System

Dominant approaches to grading in US higher education are ideologically grounded in the notion of meritocracy: that equality of opportunity is possible, irrespective of existing social and economic inequality. When companies like Proctorio promise to protect "the value of certification and degree programs,"[8] they reinforce the notion that assessment in higher education is fair, so long as cheating is prevented or punished. And yet, conventional grading systems are largely constituted around the reproduction of social stratification. As education researchers like Ann Gibson Winfield have shown, assessment in education is rife with unfairness. According to Winfield, the modern education system has never sufficiently extricated itself from its roots in eugenicist ideology and its racialized scientism around assessment.[9] Eugenicist ideology was pervasive across the first three decades of the twentieth century and has contributed to dominant ideas about how to sort and classify student achievement ever since. Today's higher education grading systems are part of a long tradition of excluding environmental factors to explain observed differences in educational performance. Throughout the early twentieth century, there were efforts to produce seemingly objective and standardized sets of educational records to engage in oppressive social sorting and render student performance legible to other institutions, including prospective employers.

As K–12 and higher education institutions became more systematized and socially interconnected to one another and the job market, and as student-teacher relationships became increasingly less intimate because of growing class sizes, grading offered a veneer of objectivity. The consequences of this have been the systematic exclusion and marginalization of low-income students and students of color in US higher education, which grading systems mask as an unfortunate but fair outcome of a purportedly meritocratic system assessing individual deservingness. However, the rise of standardized grades has not been without its critics. In the 1960s, a number of scholars, educators, and educational activists argued that grades do not promote learning, that conventional grading systems are highly racially and economically biased and unreliable, and that grades encourage unhealthy levels of competition and anxiety in students.[10] More recently, scholars have argued that grading systems in higher education are best understood as surveillance technologies that help make possible forms of extraction, discipline, and race-class sorting in higher education while shaping students' perceptions and experiences of punishment in ways that align with carceral institutions.[11] Grades can impact a given student's access to courses, scholarships, and academic services, as well as their loan repayment and visa status. Alternatives to conventional grading like ungrading, where students self-assess their work through ongoing reflection and dialogue with their instructors, do not typically lend themselves to the systems of standardization generally used to make decisions about students' relative worthiness or aptitude. Behind efforts of grade standardization and objectivity have always been, in practice, the reproduction of structural inequality and an acceptance of imposed institutional constraints.

To further illustrate this point, we can think of the switch from narrative evaluations to letter grades at the University of California, Santa Cruz, in February 2000. UC Santa Cruz was founded in 1965 to explicitly offer an alternative to impersonal, highly bureaucratic universities. The argument in favor of this grading policy change was not only one that naturalized a politics of austerity—the idea being that class sizes had grown so much, it was impossible to get to know students well enough to write good evaluations—but also one premised on needing to better exclude students with lower SAT scores, a form of standardized assessment with well-known racial and economic bias.[12] Furthermore, for four years leading up to this decision in 2000, policy was passed in California to exterminate affirmative action initiatives, which dramatically shifted the racial and socioeconomic demographics of the higher education student body away from diversification.[13]

Assumptions about "normal," compliant test-taking behavior baked into tools like Proctorio are a contemporary manifestation of the history of white supremacy embedded in college grading in the United States. While Proctorio's website stresses that, ultimately, it is instructors who make the final determination about whether a given student's flagged behavior is indeed cheating, this position indicates the limitations of human-in-the-loop approaches to algorithmic decision-making systems more broadly. Within dominant algorithmic fairness, accountability, and transparency frameworks, making sure a given algorithmically informed decision-making tool involves both automated and human judgment is often assumed to help ensure these decisions are fair and accurate.[14] And yet, the problem with exam proctoring software is not simply that it gives algorithmic tools power in determining whether a behavior is suspicious or not, but also that it is embedded with socially constructed assumptions about what grades measure in the first place. Thus, ensuring that it is ultimately a human instructor that decides whether an algorithmic flag of cheating is accurate sidesteps the issue of whether test proctoring should be privatized at all. More fundamentally, educational assessment is already riddled with unfairness. This unfairness manifests in which students receive sufficient K–12 preparation and faculty mentorship, which students have time that isn't taken up by family or work responsibilities to prepare for exams, and which students are having to overwork themselves to combat racist or ableist assumptions about their aptitude or the privileging of English monolingual ideas and expression.[15]

Ultimately, efforts to secure the integrity of the grading system double as efforts to secure the university's economic value in the higher education marketplace, which hinges on the notion of grades as a purportedly neutral measure of a given student's worthiness for social mobility. This is also why it is insufficient to simply improve the accuracy of Proctorio across students from different racial backgrounds or to ensure that all students have equal access to a private space with reliable internet access for test taking. Although initially, improving the technology's accessibility or its accuracy for students of color and neurodivergent students might seem like reasonable solutions, such approaches overlook how privatized education services are consolidating power over higher education. Scholars critical of AI ethics discourse have considered similar efforts to improve the accuracy of facial recognition technology (FRT) a harmful form of inclusion, wherein attempts to debias FRT can be predicated on unethical data collection practices or expose marginalized groups to increased repressive surveillance.[16] Building on the work of sociologists Louise Seamster and Raphaël Charron Chénier, the improvement of automated exam proctoring software or FRT can be considered examples of "predatory inclusion," in that these improvements are presented as expanding access and opportunity to marginalized people while, in practice, intensifying asymmetries of power through surveillance.[17]

Despite documented harms, mechanisms for automating assessment are highly attractive to higher education administrators in the context of the austerity-driven expansion of class sizes and the proliferation of online courses and programs, transformations that have raised concerns about the reduction of educational quality. Test proctoring companies promise to secure the academic integrity of examinations in the wake of an increasingly online and globalized system of college education. As data privacy researcher Shea Swauger explains, in an article Proctorio attempted and failed to have retracted, this promise is part of a long-standing history of universities' discriminatory fears that diversifying the academy poses a fundamental threat to institutional integrity.[18] The rise of exam proctoring software is just one example of how universities are using digital surveillance tools to secure themselves against perceived threats in ways that ultimately render marginalized students differentially vulnerable to suspicion and exclusion.

Securing the Student

While Proctorio has attempted to distance its algorithmic proctoring technology from facial recognition technology, arguing that unlike FRT, Proctorio's facial and gaze detection technology does not involve the verification of identity, in both cases, these technologies are often used to mark individuals as worthy of suspicion based on the degree to which their behavior or appearance deviates from a predetermined standard of acceptability or belonging. Furthermore, in the case of the company Proctortrack, the idea for their online proctoring technology emerged directly out of a Transportation Security Administration project that involved searching video footage for "facial expressions deemed abnormal."[19] Today, a range of facial recognition studies are reviving a host of debunked pseudoscientific theories that formed the scientific basis for systems of economic and racial oppression in the nineteenth century. For instance, while eugenicists like Cesare Lombroso and Francis Galton argued that facial structure could be used to determine criminality, some machine learning researchers have claimed to be able to use algorithms to distinguish criminals from noncriminals based on facial images alone.[20]

In the age of the smart university, facial recognition technology is increasingly being marketed as a means for universities to secure their campuses more broadly. For instance, among a suite of tools, including surveillance cameras, video surveillance analytics for detecting and responding to "objects left behind," body cameras, and smart sensor technology for detecting vape smoke, THC, and "sound abnormalities" like shouting, the private company i-PRO offers intruder detection using facial recognition analytics. This tool can instantly notify selected authorities, administer audio alerts, and initiate door locks. According to Raja Saravanan, the principal architect of applied research for the higher education technology firm Ellucian, "On college campuses in the near future, facial recognition can enable on-the-spot classroom analytics based on audience reactions during a lecture, and better-than-ever security on campus." While Saravanan recognizes that facial recognition "can seem like an unnecessary and invasive monitoring tool," he seeks to assuage these concerns by arguing that this technology, when used "thoughtfully" by faculty and administrators, can "make student data more personal and powerful, without sacrificing security."[21] Much like the discourses concerning data analytics for student recruitment and retention, the notion of seamless personalization is being used to help diffuse mechanisms for mass digital surveillance throughout college campuses in the name of efficiency and security.

Columbus State University, Florida International University, Iowa State University, the University of Alabama, the University of Illinois, and the University of Wisconsin are all examples of US higher education institutions that deploy facial recognition technology on their campuses.[22] Facial recognition technology in education contexts is also used widely in China as well as in the UK. Its use has been proposed for monitoring student attendance and granting building access, as well as for contact tracing during the COVID-19 pandemic. According to Saravanan, facial recognition systems minimize the effort students otherwise exert when it comes to accessing a dorm room or being accounted for in a large lecture hall. Saravanan also proposes that Ellician's future products will help professors deliver more engaging and exciting lectures tailored toward their students' "emotion levels" using facial recognition technology that includes emotion recognition.[23]

Similar to higher education administrators Jay W. Goff and Christopher M. Shaffer, who argue that demands for digitally tailored recruitment initiatives are coming from the students themselves,[24] Ellucian also asserts that students "expect a universal, connected technology experience on campus."[25] According to the company's blog, the "delivery of a seamless technology experience has become an expectation—similar to the technology students encounter with other modern, consumer-like experiences—and they are taking this all into account in making their decision on which school to attend."[26] In an increasingly competitive US market, public university administrators and private vendors argue that mirroring and adopting technology industry practices throughout the higher education experience will help universities gain an edge against their competitors.

According to some estimates, the AI emotion recognition industry is expected to surpass $40.5 billion by 2030.[27] Outside of the higher education context, companies that market emotion recognition argue that these tools support a range of services, including helping employers identify strong employee candidates with "grit," resilience, and positive attitudes; aiding doctors and nurses in monitoring patient pain levels; helping retailers monitor shoppers' in-store experiences; and improving human-robot interactions by making robots more sensitive to how humans are feeling. AI emotion recognition is often used without consent, particularly in contexts that are already considered public or semipublic. Furthermore, studies have found racial bias in a variety of well-known emotion recognition tools, including Face++. In one study that compared one hundred relatively standardized images of Black and white basketball players, the Face++ AI tool rated Black player's faces on average as angrier and unhappier than white faces. These tools often fail to account for context as well as cultural, situational, and personal variance, reflecting generalizations and assumptions about emotions that are not universally applicable.[28] AI for emotion recognition is then best understood as measuring facial movements, rather than emotions themselves. Yet these tools are actively being promoted for emotion recognition in a range of contexts, including retail, workplaces, police surveillance, and education. This is particularly concerning given the propensity for Black students in the United States to have white teachers incorrectly read their affect as angry; to experience more suspensions, expulsions, and disciplinary actions than white students for the same behavior; and to experience violence from university and local police based on racialized, subjective interpretations of threat.[29]

The deployment of facial recognition technology on US college campuses has met organized opposition from students, faculty, digital rights groups like Fight for the Future, the American Civil Liberties Union, and Students for Sensible Drug Policy. This opposition has emphasized research findings documenting this technology's significant inaccuracies for people of color and concerns about its use for suppressing student participation in political protests, as well as for carrying out US Immigration and Customs Enforcement deportations more broadly.[30] According to Claire Galligan, Hannah Rosenfeld, Molly Kleinman, and Shobita Parthasarathy's 2020 study, the use of facial recognition technology in educational settings is likely to exacerbate racism, normalize surveillance, and erode privacy.[31] For these reasons, they recommend a complete ban on the use of this technology in schools. New York became the first state to temporarily ban facial recognition software in public and private schools, but most local, state, and federal officials have yet to follow suit.

Facial recognition technology is not the only biometric technology being used at public universities, nor is it the first one. As Abigail Boggs explains, between 2001 and 2003, new US government visa policies were instituted that included the Student and Exchange Visitor Information System and the 2002 implementation of the National Security Entry-Exit Registration System, which vastly expanded biometrics requirements for international students after 9/11.[32] Under SEVIS, the state uses this technology and corresponding federal processes of policing to identify, manage, and track international students, processes that universities are required to participate in by creating and maintaining highly detailed records.[33]

More recently, a few universities, such as the University of Georgia (UGA), have instituted iris scanners in dining courts. Bryan Varin, the executive director of dining services at UGA argued in 2017 that this technology would be faster, more convenient, and more sanitary than the university's previous methods of hand scans and card swipes.[34] The dual concerns of sanitation and security were also referenced as grounds for the introduction of facial recognition technology at the University of Southern California (USC), a private institution, in the wake of COVID-19. Prior to this, USC used fingerprint identification in its residence halls, beginning in 2015, to regulate dormitory access. In each of these cases, promises of increased efficiency, security, and well-being are used to make biometric surveillance technologies a seemingly mundane and beneficial part of daily life. According to an official USC statement, these forms of biometric technology are more secure than student ID cards, which can be stolen or otherwise used improperly. While director of USC housing and auxiliary services Chris Ponsiglione maintains that the university's tools do not store facial recognition and fingerprint data, USC associate professor of electrical and computer engineering Wael Abd-Almageed countered that it is impossible for facial authentication to work without making a comparison to a stored version of one's image.[35]

The question of what data about students universities collect and store is often framed in terms of securing the data itself from unauthorized parties. Certainly, there are reasons to be concerned about what happens if a student's personal information is leaked or hacked—universities have been the sites of several high-profile data breaches.[36] However, concerns about the security of student data tend to sidestep more fundamental ethical questions, such as whether this data should be collected in the first place, and whether data can meaningfully address the problems that are used to justify its collection and use.

It is notable that the education technology industry has played a significant role in shaping how universities are conceiving of and assessing the risks of partnering with third-party vendors. For instance, the Higher Education Community Vendor Assessment Toolkit, originally developed in 2016, is used in more than one hundred universities for measuring vendor risk by confirming the policies in place to protect sensitive institutional and personally identifiable information.[37] The Higher Education Information Security Council (HEISC) put this toolkit forward in collaboration with Internet2, a digital technology company with a target market that includes higher education, research institutes, and government entities. HEISC is part of the nonprofit association EDUCAUSE, whose stated mission is to encourage the diffusion and adoption of information technology for educational purposes.[38] Its roots as an organization can be traced back to the annual College and University Machine Records Conference of 1962, where twenty-two data processing directors from American universities and colleges organized, with sponsorship from IBM, to share information about how they were putting to use the IBM 1401, a computer for processing punch card data. This group, which they named the College and University Systems Exchange (CAUSE), partnered with the W. K. Kellogg Foundation–funded higher education professional organization Educom to form EDUCAUSE in 1998.[39] For IBM, sponsoring higher education meetings was a means of cultivating new markets for its products while tying its brand image to higher education. This practice continues today, with a range of corporations sponsoring conferences and offering grant funds to higher education researchers and IT professionals.

Furthermore, the question of *how* a given higher education problem is conceptualized directly shapes the kinds of solutions that receive traction at a given institution. For instance, if one of the main sources of student vulnerability to harm on US college campuses is understood not in terms of internal violence against marginalized students but, rather, of outside "intruders" on campus, then digital technologies that promise to secure the university's borders of belonging are taken up as commonsense solutions to this problem. This notion of university security mirrors the history of racialized national security surveillance in North America, where notions of otherness have been integral to justifying US empire, from the settler-colonial period through the neoliberal era.[40]

In cases where universities acknowledge sexual violence on campuses, concerns about university risk and liability generally take precedence over the needs and priorities of harmed students and faculty. Furthermore, the chancellor of UC Davis used the specter of sexual violence on college campuses to justify deploying police to break up a peaceful student Occupy demonstration against rising tuition and state divestment in 2011, arguing that such protests might attract "non-affiliates" from outside the university, putting at risk "very young girls."[41] At Arizona State University, administrators gave rape whistles to female students at their downtown campus, even though no sexual assaults had been reported there. As urbanist, historian, and cultural critic Davarian Baldwin explains, "By depicting the world beyond campus as dangerous, ASU turned students away from the city and made them a captive market for the retailers and restaurants affiliated with the university."[42] Both cases demonstrate how university administrations weaponize anxieties around campus security, particularly concerning the safety of young women, to exercise control over students.

Campus security discourses have revolved significantly around issues of sexual assault beginning in the late 1980s after the rape and murder of Jeanne Clery, a white student at Lehigh University, by Joseph Henry, a Black student there, in 1986. Prior to the late 1980s, and as early as the seventeenth century, murders, lynching, rapes, violent assaults, vandalism, and riots were not uncommon on college campuses.[43] However, Clery's murder catalyzed large-scale mass media and political mobilization efforts among members of the public, advocacy groups, and researchers and helped position universities as besieged by crisis due to purportedly new and rising threat levels to innocent students.

Clery's case was unrepresentative of campus crime at the time; most was nonviolent, and cases of sexual assault usually involved people who knew each other and were of the same race, as is true today. Nonetheless, the case played a significant role in shaping the rise of the 1990 Campus Security Act. This act made statistics the dominant mode of institutional reporting about campus sexual violence.[44] As sociologists Rebecca Dolinsky Graham and Amanda Konradi explain in their close reading of journalistic accounts of the period, the Clery case received immense media attention largely because of the racial identities of the students involved. Media accounts typically framed Henry as "an outsider to the campus mystique, although he was a student... The 'normally placid' campus life of Lehigh specifically and national campus life in general, was being threatened by a 'dark side' that comes with student diversity."[45] This case helped to entrench the notion that campus security issues are often caused by outsiders racialized as nonwhite, a status that can be ascribed to those both within and outside of formal membership in the academic community.

The legal consequences of this case, along with others that have found universities negligent for failing to address "known risks" to students' safety, have put pressure on universities to document various crimes and offenses, electronically warn students, and take up additional measures for security. These measures have included a range of technologies, such as key cards, video surveillance, and crime-mapping technology for analyzing "criminal victimization patterns" and "making better use of personnel."[46] Smart technologies for campus policing purportedly provide effective, cost-efficient, and diffuse networks of security to meet reporting requirements and mitigate crime. The changing social and legal context of the university has coincided with the use of police, and corresponding policing technologies, for preventing victimizations that universities can be held financially liable for.[47] In several cases, beyond universities mirroring many of the assumptions and tactics that underpin racialized policing in the United States, the companies providing these technologies also directly supply and maintain technology used by federal government and local law enforcement agencies. For instance, the website for UBTOS USA Inc., a South Korean firm that manufactures facial recognition devices, boasted of contracts with several Southern California school districts, hotels, and the Los Angeles Police and Sheriff's Departments.[48]

Both campus and municipal police departments have faced nationwide protests over issues of racial profiling and violence in the wake of the Black Lives Matter movement, with similar calls for defunding, disarming, or abolishing campus police departments after decades of unsuccessful modest reforms for reducing police violence.[49] Public universities have received a range of military technologies through the Department of Defense since 1998, including 12-gauge shotguns, Kevlar helmets, an M-79 grenade launcher, and M16 assault rifles. Ninety-two percent of four-year public universities with more than 2,500 students currently have an armed police force, and students and faculty of color consistently report experiencing greater levels of surveillance and policing, suffering declines in mental health as a result of police encounters, and being at significantly greater risk of being killed by police.[50] Thus, although digital technologies like AI systems trained to recognize the sound of gunfire or the image of a gun are marketed as an objective means of securing campuses against gun violence, these systems can falsely identify objects and sounds that put people of color at significantly greater risk for police violence.[51]

Ultimately, it is impossible to abstract smart university security technologies from the role of policing as an institution that helps to sort, classify, and control Black, brown, poor, and dissident populations.[52] These technologies provide seemingly neutral and objective approaches to crime prevention that entrench the racial and economic inequalities that pervade the academy and its relationship to surrounding communities. For example, with gunshot detection technology, which uses acoustic sensors trained to detect the sound of gunshots, some universities have acquired this technology from private vendors, while others access the technology through direct partnerships with law enforcement agencies. Public universities like the University of Nevada, Reno, have paid hundreds of thousands of dollars to install gunshot detection technology well beyond the immediate boundaries of the university's main campus into nearby poor communities of color, which local police departments control.[53] Rather than helping to address the underlying structural drivers of gun violence in surrounding communities that include restrictions on the accessibility of education and economic opportunity, universities often exacerbate these problems by intensifying forces of gentrification, criminalization, and exclusion. As Baldwin's work has demonstrated, police are a key mechanism through which universities carry out expansion efforts and secure control over public land, including through intensive surveillance of people of color's behavior in areas where universities plan to displace local communities.[54]

Direct data sharing between universities and police departments also happens through digital technologies like automated license plate readers, which are marketed as tools for making parking enforcement more efficient while doubling as public safety measures.[55] Unlike GPS devices, automated license-plate readers are not considered a violation of US citizens' privacy rights, making these tools particularly attractive to police departments as well as the US Department of Homeland Security. Furthermore, some digital technologies have been used for the shared interests of university administrators and law enforcement in suppressing organized political action on college campuses. One such example is Social Sentinel's software for social media monitoring, which was used to watch student activists at the University of North Carolina protesting the Confederate memorial, Silent Sam. Social Sentinel uses an AI-based language engine to scan digital content, including social media, institution-owned data, and sanctioned emails, for signals regarding "safety and security threats."[56] The company then uses machine learning to classify whether signals should be treated as actionable to guide selected decision makers. The use of this technology on student protestors speaks to the degree to which the notion of a "safety and security threat" involves high levels of discretionary power, even as these digital tools are presented as more objective and empirical mechanisms for security-related decision-making.

While a rise in mass shootings at college campuses in the past fifteen years has been cited to justify acquisition of surveillance technologies more commonly used for metropolitan policing,[57] universities have deep historical ties to law enforcement for exercising social control over college campuses. In the late nineteenth and early twentieth centuries, campus security typically consisted of a few designated "watchmen" whose primary duties were custodial, such as protecting college property and detecting fire hazards. Yale University established the first campus police force in 1894, hiring two New Haven police officers to patrol its grounds in response to rising tensions between community members and the Yale Medical School. During the 1950s, as universities began a period of unprecedented growth in student enrollment and physical size, universities also hired retired municipal police officers to serve as directors of campus security. These directors sought to actively construct campus security in the image of urban police departments, separating their duties from custodial work and creating formalized departments with designated officers instead. Campus officers also began meeting for conferences, standardizing their practices, and strategizing for ways to increase their power during this period.[58]

Owing to the doctrine of in loco parentis—which conceptualized universities as stand-ins for students' parents during their formal education and gave them the power to regulate and punish student behavior—security officers could detain outsiders for arrest, but students and faculty would receive university discipline. In loco parentis also meant that American universities, both public and private, were free to impose and strictly enforce "character-building rules" that included restrictions on student speech, socialization, and movement. This doctrine also permitted universities to summarily expel students without notification of charges or providing evidence, often with disparate consequences for students at women's colleges and historically Black universities.[59]

It wasn't until the late 1960s and early 1970s that courts gradually abandoned in loco parentis so that higher education could respond to challenges presented by the integration of historically marginalized groups after the passage of *Brown v. Board of Education* and the Higher Education Reauthorization Act of 1965.[60] Lacking a sufficient mechanism of social control to suppress rising anti-war and civil rights movements on college campuses, police agencies and the National Guard were brought in to suppress protests. In the infamous 1970 anti-war protests at Kent State University, the Ohio National Guard killed four students and wounded nine others. Just eleven days later, local police officers killed two students and wounded twelve others at Jackson State College (now University) in Mississippi and South Carolina State, two historically Black colleges. By the early 1970s, thousands of students were facing criminal charges for participating in campus marches, sit-ins, draft-card burnings, and protests.[61]

In an effort to regain control over campus security during this period, universities successfully lobbied for legislative approval to create their own fully fledged internal police agencies, often with little accountability or oversight besides that of the president of the university or a vice president of student affairs.[62] Campus police forces became larger, received more extensive legal powers, and expanded their acquisition of weapons. At the University of California, Los Angeles, in a partnership between the university and city police in 1972, undercover police agents, registered as students, infiltrated classrooms and university-registered organizations, and generated police dossiers on faculty and students. Professor Hayden White sued the LAPD chief of police and won on First Amendment grounds in 1975.[63]

Yet the use of police on campus in the name of national security only intensified further following the September 11 attacks of 2001. A range of scholars and journalists have argued that these attacks were strategically leveraged to justify and expand the power of US empire in the Middle East and curtail civil liberties at home, with particularly significant consequences for political dissidents and people of Middle Eastern and North African descent.[64] Campus police departments worked with members of the FBI, provided access to Middle Eastern students, and in some cases, received federal security clearance and were tasked with surveilling students as part of the FBI's Joint Terrorism Task Forces with local law enforcement.[65] At some campuses like the University of Arizona, campus police have been called to arrest students protesting border agents on campus and have partnered with city police to repress off-campus anti-ICE protests.[66]

Occupy movement protests at universities in 2011 sought to challenge the power of private capital over university life, state divestment in public services like education and healthcare, tuition hikes, and the impoverishment more broadly of everyday people; these nonviolent student protestors were met with violent, militarized police repression, much like their counterparts off campus. As just one example, the University of California, Davis, spent at least $175,000 to bury online search references to a 2011 incident of campus police brutality where an office pepper-sprayed seated Occupy protesters at close range.[67] In reflecting on these protests, professor of political science Farah Godrej argued that "the administrative response to campus protests has managed to completely subvert the logic of nonviolent protest, effectively criminalizing all forms of it by focusing on the potentially threatening nature of such protest."[68] Such tactics significantly enable violent repression of students' political speech.

This idea of potential "threats" has also been used to justify the diffusion of a range of advanced digital technologies for policing at on-campus protests, from the deployment of drones at UC Berkeley ahead of demonstrations against conservative campus speakers, to allegations of facial recognition technology being used to track down students at the University of Miami demanding better COVID-19 protections and sick pay for university contract workers.[69] Furthermore, during the 2019–2020 graduate worker wildcat strike at the University of California, Santa Cruz, for a cost of living adjustment (COLA) in response to exorbitant housing and living costs, university administrators turned the campus's learning management system and the videoconferencing platform Zoom into strikebreaking mechanisms. The university had directly contributed to these conditions by expanding undergraduate enrollments that consolidate the housing market; by burdening graduate students with debt, which they take on to compensate for their low wage work for the university; and by taking no meaningful action to address the market pressures of nearby Silicon Valley on housing availability and costs. Receiving little administrative movement, what began as a grading strike, with 351 graduate workers withholding their grades from Canvas, became a campus picket and full work stoppage.[70]

Administration encouraged faculty, graduate students, and lecturers to use Zoom to teach classes if necessary and asked students to report on faculty and graduate students who canceled, moved, or repurposed their classes to support the living-wage campaign. Warnings from UC president and former secretary of homeland security Janet Napolitano, as well as from the International Student and Scholar Services, were issued to intimidate students. These warnings threatened disciplinary action and employment termination for strike activity, which would mean for international students the loss of tuition waivers and the invalidation of their visas. A militarized police force costing approximately $300,000 a day and equipped with military surveillance technology was used to further crush the picket, whose numbers had expanded considerably owing to the efforts of graduate-undergraduate solidarity organizing led by Black, Latine, and undocumented students.[71] These students saw the strike as a means of mobilizing against conditions of economic precarity as well as racialized policing in the university.

By the time the university transitioned to virtual operations in response to the pandemic and shelter-in-place orders, a variety of repressive tactics had been deployed, including police arrests and beatings, threats of multiyear suspensions and deportations, and a loss of campus housing and employment, all of which the university justified in the name of preserving the safety and security of campus for students. Furthermore, the administration linked the security of students to the security of the grading system itself. In the words of Napolitano, graduate workers were holding undergraduate grades "hostage."[72] This move to anthropomorphize the grading system was an effort to not only claim grades as university property but also to whip up support for the university's repressive tactics among undergraduate students, whom the university repeatedly framed, following the analysis of UC Santa Cruz graduate students Gabe Evans and Taylor Wondergem and faculty member Nick Mitchell, as "vulnerable, harmed, ripped-off consumers."[73] The grade, which leads to the conferral of the degree, is what promises to offer social mobility under capitalism.

Additionally, the idea that the university is like a city, and therefore vulnerable to issues of congestion, crowding, and crime, is often used to rationalize the university's diffusion of the same policing techniques that currently dominate municipalities. According to proponents of smart technology for campus security, this security surveillance infrastructure doubles as a means for optimizing campus-wide traffic, facility maintenance, energy usage, and staff management.[74] It is also the case that campus police chiefs have sought to gain legitimacy by not only modeling municipal policing but also actively establishing themselves, in the words of scholar Grace Watkins, as "leading experts on campus security within a global network of carceral actors."[75] Some US university police forces have participated in off-site specialized training and networking with Israeli police, intelligence experts, and military officials whose tactics are used to occupy Palestinian territories, and have partnered with international police departments through training programs that "stress the superiority of American policing techniques."[76] This more recent effort is part of a long-standing historical practice wherein US police forces have taught policing techniques to a range of countries to secure US economic and geopolitical interests, while bringing back techniques acquired abroad to implement on marginalized and dissident populations at home. This historical practice has included importing the surveillance and repression techniques used to combat resistance to US colonization during the Philippine-American War, as well as teaching techniques to police forces abroad in countries with communist insurgencies during the Cold War.[77]

Campus police have also sought legitimacy through direct partnerships with city police and through the International Association of Campus Law Enforcement Administrators (IACLEA) partnerships with a range of digital technology providers and corporate sponsors, including Axon, Vector Solutions, and Uber. The IACLEA promotes these firms' products and services in exchange for funding. Originally founded in 1958 as the National Association of College and University

Security Directors, the IACLEA remains the largest lobbying and professional organization for campus police in the world. The IACLEA's partners also include private security firms that have run private prisons and detention centers for adults and children, provided staff for Guantanamo Bay, and transported detained people in partnership with ICE.[78]

It is through these partnerships and advocacy efforts that campus police have worked to resist fiscal austerity for themselves, while repressing social movements against austerity, like Occupy and COLA, on college campuses.[79] While public funding support for education continues to decline, many campus police departments have experienced a steady increase of spending. Thus, the increasing militarization and repressive mobilization of campus police, including the strategic acquisition and deployment of smart technologies for campus security, has developed hand in hand with rising austerity measures at public universities. The enforcement of austerity politics in US public education necessarily requires the use of police to repress and criminalize dissent in ways that are institutionally legitimized.[80]

Furthermore, campus security is inextricably tethered to the larger context surrounding US policing. As American studies scholar Dylan Rodríguez explains, "Campus police and their administrators already view themselves within a global and historical context of managing, containing, co-opting, and repressing the potentially explosive and disordering political mobilizations that can happen just outside their office windows."[81] Universities have attempted to maintain the status quo in response to these global historical conditions not only through direct police repression on college campuses but also by taking an active role in producing the very forms of scientific knowledge and technical practices needed for preserving hegemonic power structures in the United States.

Securing the Nation

Abigail Boggs and Nick Mitchell have shown that even within critical university studies scholarship, there is often an assumption that the university is, at its core, progressive, and that its "drive to know the world is virtuous."[82] Yet dominant scientific and technical epistemologies in university research are deeply tied to histories of racism, colonialism, militarism, capitalism, and male supremacy.[83] Although scholarship typically periodizes the formation of the academic-military-industrial complex beginning in World War II, we can trace the university development of contemporary carceral technologies like predictive policing back to earlier forms of university-sponsored techno-scientific racism, where university research helped justify the dispossession of Native people and the enslavement of Black people. As Boggs and Mitchell explain, colleges "reproduced slavery to the extent that they served as pivotal sites in the production, legitimation, and dissemination of dominant ideas for emerging generations of the colonial elite . . . Academic institutions became the site of alchemical transformation of highly biased and interested observers of African and Indigenous populations into legitimate knowledge."[84] Historian Craig Steven Wilder's work documents how the organization of science faculties and medical colleges in the colonies coincided with slave owners, planters, land speculators, and Atlantic merchants funding scientific research, contributing to the production of scientific racism as a "legitimate" form of knowledge about human differences and the supposed biological supremacy of white people.[85]

People of color were frequently subject to invasive and involuntary research in both life and death throughout the eighteenth century for the edification of American and European medical students. In the nineteenth century, the creation of federal land-grant colleges for advancing agricultural and mechanical education as well as instruction in military tactics was predicated on expropriated Indigenous land acquired through coercive treaties and outright land seizures. These acts of dispossession coincided with academic discourses that framed Native people as incapable of civilization because of fixed biological traits while deploying ethnography, philology, and geography to map Native languages, mineral resources, and the geologic structure of Native lands for the project of US settler colonialism.[86]

With the US rise as a global superpower during World War II, President Franklin D. Roosevelt issued an executive order to create the Office of Scientific Research and Development, which decisively aligned university research efforts of the period with militaristic national policy priorities. Universities would go on to support the development of missile technology, gunsights, bombsights, radars, chemical and biological weapons, and the atom bomb for the war effort.[87] Under McCarthyism, public universities' state funding shifted from land-grant agricultural resources to the defense establishment, which further laid the groundwork for "deep strata of connection and complicity between imperial statecraft and the knowledge complex of the U.S. academy."[88] Following the launch of *Sputnik*, a US Department of Defense directive created the Advanced Research Project Agency in 1958, ultimately known as the Defense Advanced Research Projects Agency. DARPA led to the rise of computer science departments and the internet protocol ARPANET, which produced the internet. There was broad consensus and collaboration among many scientists, academics, and the military grounded in shared notions of US security needs.[89]

By 1980, the Department of Defense as well as the State Department were likely to insist on limiting access to state-funded university research, arguing that transparency about new technology could jeopardize the nation's economic standing and, by proxy, national security.[90] This shift coincided with the rise of the technology-transfer movement, which pushed to commercialize academic research through private sector licensing, rather than making federally funded research freely available to the public. Since 2008, several institutions of higher learning have become home to corporate-endowed data science institutes, centers, and programs that have contributed to a suite of digital technologies—including AI and drones—used for surveillance and killing in Pakistan, Yemen, and Somalia.[91] Today, the Department of Defense is playing an increasingly large role in funding research that was previously supported by the National Science Foundation. Science, technology, engineering, and mathematics program partnerships with defense contractors like Lockheed Martin are promising to alleviate student debt through scholarships and job placement initiatives that often explicitly target students from minoritized groups in the name of diversifying STEM fields. Furthermore, several contemporary partnerships between the academy, the US government, and private businesses have resulted in the proliferation of whitewashed technology ethics discourses that are strategically aligned with the priorities of both the US military and Silicon Valley.[92] At public universities, these partnerships can make it possible for companies to benefit from profitable research while evading taxation, taking advantage of these universities' tax-exempt status.

Some argue that universities have an institutional responsibility to contribute to national security through the advancement of digital technology. For instance, according to political scientist Amy Zegart and CIA analyst Michael Morell, the rise of AI, biotechnology, and quantum computing have equipped US adversaries with technological capabilities that erode traditional US intelligence advantages. They argue that bridging the divide between the technology industry and the intelligence community is a national security imperative requiring "input from technology companies, civil society, and academia."[93] In 2020, the National Security Commission on Artificial Intelligence, spearheaded by Big Tech executives, called for increased US government funding for national AI research infrastructure. The National Science Foundation has partnered with Amazon to support research on fairness in AI to promote the diffusion and adoption of the technology across a range of sectors in ways that can significantly undermine academic freedom.[94]

The remainder of this chapter looks at examples of contemporary university research on improving three types of AI technologies that are used to criminalize and punish in the United States: facial recognition technology, Correctional Offender Management Profiling for Alternative Sanctions, and predictive policing software. These efforts are considered not in isolation but rather as forms of knowledge production that are symptomatic of the university's broader historical alignment with the security establishment, on the one hand, and with contemporary corporatization measures, on the other. Both Big Tech and the US government hold an inordinate amount of power over the direction of university AI research, given modern AI's dependency on significant financial resources to support data- and computing-intensive techniques. As AI researcher Meredith Whittaker explains, industry and military capture of academic research in AI has "perilous implications for academic freedom and knowledge production capable of holding power to account."[95] Yet these partnerships are highly attractive to many university administrators and researchers, given the financial and reputational advantages they afford.

There are also significant professional risks in contesting these research initiatives for precarious students and faculty.[96]

Facial Recognition Technology

Between 2012 and 2014, researchers at Duke University, Stanford University, and the University of Colorado at Colorado Springs (UCCS) captured images of students traveling between classes and visiting on-campus cafés. These images were taken to help improve facial recognition technology's capacity to recognize people in motion in a "natural" environment.[97] At UCCS, in a study approved by the Institutional Review Board (IRB), long-range surveillance images of over 1,700 people were collected from February 2012 to September 2013 without students' knowledge or consent, which helped to produce a dataset with desired real-world conditions— poor lighting and partially obstructed views—for defense and intelligence agencies.[98] US defense and intelligence agencies funded this study, including the Intelligence Advanced Research Projects Activity, the Office of the Director of National Intelligence, the Office of Naval Research, the Department of Defense Multidisciplinary University Research Initiative, the Special Operations Command, and Small Business Innovation Research.

Several of the lead researchers on these projects have defended their work by arguing that these datasets will ultimately result in more accurate AI. Existing forms of facial recognition technology have been shown time and again to have higher error rates for women of color and transgender people.[99] However, this rationale ultimately justifies widening surveillance's net and, by proxy, the reach of systems of oppression that deploy automated recognition. Simone Browne's work has demonstrated the long historical arc of the relationship between surveillance and racial oppression, from the transatlantic slave trade to contemporary forms of racial profiling.[100] More accurate facial recognition will not rectify a criminal legal system that is inherently unfair, given the ways histories of slavery and segregation, as well as ongoing race- and class-based oppressions, shape processes of criminalization, resulting in unjust treatment for communities of color and cycles of poverty and incarceration.

Surveillance is often bound up with the monitoring and disciplining of boundaries that sustain racism, patriarchy, and capitalist exploitation.[101] Efforts to make marginalized groups more legible to these systems illustrate the ways ideas of inclusion, equity, and identity have been co-opted for the purposes of improving mechanisms for social control.[102] Furthermore, IRBs are limited in their capacity to address the power relationships that condition who gets to conduct university research and how this research impacts the rights and well-being of marginalized people. As communication scholars Jasmine R. Lindabary and Danielle J. Corple have noted, "In the context of academic research, power is ... nestled in institutions and structures (e.g., IRBs) that define what constitutes privacy and dictate whose vulnerabilities count."[103] Yet these facial recognition studies were defended on the grounds of having received IRB approval, which federal regulations require for research involving human subjects.

The inadequacy of IRBs for meaningfully addressing privacy rights in the context of machine learning research is particularly acute given that the IRB focuses on whether individuals are identified, and if so, whether being identified puts them at risk of financial or medical harm. Regardless of whether individuals in these studies are personally identified, their images are being used to train technologies that will be used to monitor, criminalize, and punish marginalized and dissident populations when put in the service of institutions like policing and border security. In both the case of Proctorio and these university facial recognition studies, student information becomes fodder for improving harmful, repressive tech. However, students are certainly not the only group whose images have been used to test facial recognition technology without their knowledge or consent. This list also includes immigrants, abused children, people who have had mug shots taken, and deceased people.[104]

Over the past thirty years in the United States, university research ethics discourses have largely shifted from questions of what morals or rules should guide research conduct to questions of what will help mitigate concerns about university liability.[105] In these studies, one liability concern would be whether students have a reasonable expectation of privacy in the areas of campus researchers intended on photographing. Yet a range of scholars have demonstrated that traditional understandings of privacy typically do not account for the ways privacy intersects with relations of power and inequality.[106] This traditional notion of a reasonable expectation of privacy originally comes from the Supreme Court decision in *Katz v. United States* (1967), which laid the foundations for what would become the "reasonable expectation" test, based on whether the individual exhibits an actual (subjective) expectation of privacy and whether this expectation is "objectively" reasonable. In the lead researcher's defense of the UCCS study, he noted that students have no reasonable expectation to privacy when walking around campus, and that there is nothing illegal about taking photos of people in public.[107] However, although one can easily imagine being seen, overheard, or even photographed by others in the vicinity, in the UCCS study, students were photographed from more than 492 feet away without their knowledge, accelerating the advancement of multi-target multi-camera tracking and other forms of computer vision. Dominant understandings of privacy norms, which the academy institutionalizes and reproduces, fail to address how advances in digital technology intensify conditions of vulnerability to social structures of domination.

COMPAS

Another AI tool university researchers have contributed to developing is the Correctional Offender Management Profiling for Alternative Sanctions, a carceral risk assessment tool used in US courts, which provides a prediction of the likelihood of a given defendant repeating a criminal offense compared to individuals from similar data groups. In 2016, *ProPublica* researchers found that this tool disproportionately falsely labeled Black criminal defendants in the United States as being at high risk for committing future crimes.[108] The company Northpointe (now Equivant), the owner of COMPAS, countered that the tool was fair because of accuracy equity across racial groups. Ultimately, it became clear that because the COMPAS training data was embedded with past patterns of historical discrimination against Black defendants, predictive accuracy between racial groups was mathematically impossible to achieve alongside equal false positive rates.[109] This controversy gave rise to benchmarks for assessing different statistical definitions of fairness in machine learning.

Many university researchers, including those who have received corporate and government funding to improve the fairness of AI, will use COMPAS data to develop solutions for addressing algorithmic unfairness using these benchmarks. However, putting aside these debates, the COMPAS tool is itself biased toward incarceration. First, unlike other types of reforms for addressing unfairness in the criminal legal system, such as ending cash bail or pretrial detention, only the relationship between incarceration and recidivism has been measured in a manner that is conducive to machine learning.[110] Second, historical court and arrest data are not neutral but, rather, a reflection of a police system that treats people of color more harshly at every turn, from who is targeted for arrest, to how judges rule, to sentencing.[111] When university researchers improve the accuracy of COMPAS, they are improving the efficacy of a system that disproportionately polices marginalized people who are inequitably subjected to criminogenic factors in the first place. As the Coalition for Critical Technology explains, "Such research does not require intentional malice or racial prejudice on the part of the researcher. Rather, it is the expected by-product of any field which evaluates the quality of their research almost exclusively on the basis of predictive performance."[112] Nonetheless, university initiatives grounded in direct research collaborations with, or service to, policing in the United States shore up the contested legitimacy of the carceral state at the precise moment when digital technologies for racialized security are becoming more diffuse throughout daily life, both on and off campus.

Predictive Policing Software

Predictive policing software is an additional form of carceral technology that university researchers have been instrumental in helping to design, implement, and assess. Early predictive policing measures were initially developed in the early 2000s by the LAPD and have since grown to include a range of precincts, including Chicago, Memphis, Minneapolis, Dallas, and New York. Predictive policing algorithms are used to decide where to deploy police or to identify individuals who are purportedly more likely to commit or be a victim of a crime. The marketing and design of these tools deploy the same anticipatory logics as smart city discourse, in that both attempt to use rational thinking, data calculation, and control to ward off financial and social crises under austerity.[113] Furthermore, the idea of "smart cities" and predictive policing gained traction more or less around the same time: during the early 2000s economic downturn that became known as the Great Recession.[114] Most predictive policing projects are federally funded and in many cases are designed in partnership with in-house data scientists, university researchers, and private companies. Two such examples are Blue CRUSH, which developed in 2006 out of a collaboration between the Memphis Police Department, the University of Memphis, and corporations IBM and SkyCop, and PredPol, which developed out of a partnership between the Los Angeles Police Department and UCLA researchers in 2012. Both Memphis and Los Angeles have significant levels of poverty and wealth inequality, with Memphis being one of the poorest and most segregated metro areas in the nation.[115]

Proponents of predictive policing technology argue that it increases the accuracy and efficacy of policing tactics, while reducing costs and officer bias.[116] Developers and police forces argue that the more commonly used place-based predictive policing, which uses preexisting crime data to identify places and times that have a high risk of crime, allows them to focus on the "where" and "when," rather than the demographic traits of "who" regarding determining the potential of a crime to occur. As urban studies researchers Simone Tulumello and Fabio Iapaolo critically point out, "By using technology intensively, and by putting governments in partnership with private companies, predictive policing offers a future-oriented, 'technically neutral' solution to crime and violence."[117] In practice, predictive policing further entrenches existing racial injustice within the criminal legal system, resulting in disproportionate stops, searches, arrests, incarcerations, and deaths for people of color in the United States under the guise of greater objectivity and neutrality.

Given that low-income communities of color are policed at higher rates than white and affluent communities are, existing datasets tend to perpetuate discrimination in policing and heightened levels of vulnerability to state surveillance for marginalized people. In a 2016 study of PredPol's predictive policing algorithm using data derived from Oakland police records, researchers found that Black people were targeted for predictive policing of drug crimes at roughly twice the rate of white people, even though the two groups have roughly equal rates of drug use, resulting in a feedback loop that would intensify the policing of Black communities further.[118] This is true despite the fact that PredPol advocates argue that the tool is bias-free: "PredPol uses only three data points in making predictions: past type of crime, place of crime and time of crime. It uses no personal information about individuals or groups of individuals, eliminating any personal liberties and profiling concerns."[119] This example is part of what interdisciplinary scholar of science, medicine, and technology Ruha Benjamin identifies as the New Jim Code: forms of technological practice that reproduce structural racism and yet appear to offer greater impartiality.[120] Similarly, in a study of Blue CRUSH published in 2021, Tulumello and Iapaolo found no evidence of its capacity to prevent crime based on a study of official crime data, but rather, its ability to intensify conditions of surveillance for marginalized people in Memphis with little public scrutiny.[121] When university researchers uncritically engage in research that uses data from the existing criminal legal system and develop tools for it, they contribute to anti-Black police violence and to academic discourse that fuels deeply historically entrenched stereotypes about African Americans as deviant populations in the United States.[122]

Additionally, in a study from the AI Now Institute, researchers found that dirty data—data taken from corrupt, racially biased, or otherwise illegal policing practices—either directly impacted or was at risk of impacting the development of predictive policing tools.[123] Legal scholar Andrew Ferguson argues that predictive policing might make it easier for police officers to meet the legal threshold of reasonable suspicion for stopping individuals, given the use of both policing-related and noncriminal data for making predictions.[124] Some predictive policing programs have reportedly used welfare records, renter and homeowner data, and census records as data sources. For this reason, Ferguson has argued for greater transparency and management of the use of predictive policing tools. However, as sociologist and law professor Dorothy E. Roberts explains, "Without attending to the ways the new state technologies implement an unjust social order, proposed reforms that focus on making them more accurate, visible, or widespread will make oppression operate more efficiently and appear more benign."[125] The historical roots of modern US policing are found in the laws and repressive practices that were used for enforcing slavery at home and imperial power abroad and have subsequently influenced the development of policing as a key mechanism for maintaining white supremacy and racial capitalism in the present.[126]

Given the lack of transparency and accountability for predictive policing measures, as well as the technology's civil and privacy rights implications, the Stop LAPD Spying Coalition has lobbied to fully halt the use of predictive policing. At UCLA in 2019, twenty-eight professors and forty graduate students sent a letter to the Los Angeles Police Commission that rejected the empirical merit and ethics of the predictive policing research underpinning Pred Pol. The following year, the LAPD used the university's Jackie Robinson Stadium to detain protestors and others for curfew violations during an uprising after George Floyd's murder. In response, a coalition of UCLA students, faculty, and community organizations formed to work toward divestment from policing and investment in reparative public goods. Students led protests in partnership with Stop LAPD Spying against UCLA's complicity in mass incarceration and racial injustice through its research partnerships with the LAPD and deployment of university police.[127]

Rather than discrete, isolated forms of digital technology innovation happening on college campuses, I have argued that the examples featured in this chapter are part of broader university logics of securitization rooted in long-standing historical systems of racial and economic oppression. Universities not only apply some of the same disciplinary and surveillance technologies as carceral institutions, but they also actively help produce carceral tools for criminalizing and controlling marginalized populations. What would it mean to link efforts to resist racializing security technology on college campuses, the militarization of university police forces, and the university production of carceral technology as part of a broader politics of resistance against the smart university? The following chapter of this book considers the possibilities and limits of critical digital literacy training for resisting the production and incorporation of harmful digital technology on college campuses in order to imagine a different future for higher education.

Chapter Six

(Beyond) Digital Literacy

Proponents of smart universities have often referred to students born between 1995 and 2010 as "digital natives." They are, so the discourse goes, highly versed in social media, consume self-selected news online, and desire individualized, technologically mediated mentorship. These students are "radically different from those of the past," and universities must "adapt their systems to meet the unique and evolving needs" of these learners.[1] This discourse locates the integration of digital technology into higher education as a response to these students' intrinsic wants as opposed to the priorities of higher education administrators and education technology companies. There has been significant critique of the term "digital natives" that has led to recent declines in its popularity, including that it neglects varying attitudes, levels of familiarity, and access to technology among students from the same generation. Its use of "native" as a metaphor also contributes to the erasure of Indigenous people in dominant discourses about technology.[2] Nevertheless, its underpinning assumptions persist in research studies on student perceptions of technology as well as in education technology marketing literature. This literature obfuscates how students are being actively socialized according to the rhythms and demands of the digital economy in a context of rising wealth inequality and precarity. It suggests instead that universities *must* integrate digital technology into students' higher education experiences or risk failing to compete with the universities that do. However, not all students are equally convinced by the smart university's promise of convenience and efficiency or find the digitization of their universities seamless or effective. In communication researchers Pauline Hope Cheong and Pratik Nyaupane's recent study using student focus groups at a large US public university, some students expressed an inclination to trust public universities with their data, while others wanted more control, privacy, or opportunities to participate in the design of smart technology for their campuses.[3]

Nonetheless, with varying degrees of transparency, universities are increasingly integrating data-intensive digital tools into many facets of student life and learning. Education policy analysts Manuela Ekowo and Iris Palmer have emphasized the importance of university transparency about how and why they use data in order to ensure accuracy, comprehensiveness, and minimize discrimination.[4] Yet Ekowo and Palmer also acknowledge that the use of outside vendors and the legal protections afforded to private companies can make transparency difficult. As argued in chapter 3, it is also not enough to demand transparency, especially when it is not coupled with meaningful forms of accountability for private industry or democratic control over whether a tool should be used at all in contrast to alternative, and perhaps nontechnical, solutions.

Furthermore, not only have student privacy laws failed to keep pace with transformations in digital technology, but privacy is also not a sufficiently robust framework for addressing issues of structural inequality in the academy. Although improving students' abilities to access and amend incorrect records and to limit what data is collected about them would be an improvement to existing student privacy laws, as proposed by the Electronic Privacy Information Center's Student Privacy

Bill of Rights, this framework puts the onus on students to individually opt out.[5] It is difficult for students to resist digital technology they may feel compelled to use, given the power differentials that structure life in the academy. Furthermore, many of the critiques offered in this book apply to faulty tools as well as tools that are based on seemingly accurate student data but which nonetheless reinforce surveillance and austerity politics in higher education, perpetuating long-standing forms of inequity. As argued in chapter 1, data is not neutral or raw but rather a product of decisions about what needs measurement, how, and why.

This is not to say that students wouldn't be well served by critical digital literacy education and, particularly, a critical digital literacy that is geared toward the academy. By "critical digital literacy," I mean educational techniques that empower learners to analyze, critique, and transform technologies and their relationships to power and oppression. However, as digital education researcher Luciana Pangrazio cautions, it is important that such techniques account for the fact that not all learners share the same contexts, practices, or attitudes about technology.[6] One example we might use to envision the possibilities of this sort of critical digital literacy education is ChatGPT.

ChatGPT is short for Chat Generative Pre-trained Transformer, and it was launched in November 2022. Perhaps most important to stress, ChatGPT is *not* trained to generate truth. It was trained on 570 gigabytes of books, articles, websites, and social media content scraped from the internet to generate plausible-sounding statements by making predictions of what words should be paired using multiple layers of computational analysis. Higher education has been particularly consumed by questions of what ChatGPT means for issues of academic integrity, meaning what to do about the student who uses ChatGPT for their essay or exam, especially without an instructor's knowledge or permission. Quickly, a suite of private tech firms have been capitalizing on the rising AI arms race in higher ed. Already, companies like Turnitin are promising tools that can detect the likelihood that a student has used ChatGPT for a submitted assignment, despite evidence that AI plagiarism-detection tools are highly biased and inaccurate. As writing professor Whitney Gegg-Harrison stresses, the students who will be hit hardest by faulty plagiarism detection tools are the ones who are already policed the most, the students who some instructors wrongly assume can't produce clear prose: non-native speakers and speakers of marginalized dialects.[7]

OpenAI, the Microsoft-backed AI research and development firm behind ChatGPT, originally funded by prominent Silicon Valley figures including Elon Musk, Sam Altman, and Peter Thiel, has also started commercializing ChatGPT in the form of ChatGPT Plus. For twenty dollars a month, a user can gain access to the tool even during peak times as well as faster response times. Although OpenAI originally represented itself as a nonprofit committed to transparency, openness, and collaboration, the company no longer allows developers to view, assess, or build on the model, intensifying monopolization and dependency on AI companies that can rapidly scale to meet demand.[8] It also shifted to a capped for-profit company structure in 2019. OpenAI is the same company behind Whisper AI, a speech recognition model trained on 680,000 hours of audio taken from the web and which used audio from Indigenous Māori speakers without their knowledge or consent to improve its multilingual capacities. The company subsequently released and capitalized on this tool without having its quality vetted or assured by Native speakers.[9] This is just one example of how the design and development of AI tools can reproduce long-standing colonial power dynamics between settlers and Indigenous people.

The fact that the communities from which this data is taken are given no say over whether their data is used, and for what, is one of many ways we might reframe conversations around ChatGPT in the classroom to stress critical digital literacy. Students should be encouraged to question and challenge received understandings about what issues their academic community should prioritize when it comes to AI. In their effort to attract funding and talent, companies like OpenAI benefit from both positive and negative hype surrounding these tools, and they benefit from the mystification of the labor underpinning them. Recent engineer and industry leader claims that chatbots are conscious, or that we are on the precipice of AI with nonhuman minds capable of posing existential risks to humanity, distracts from the much more realistic and near-term harms that corporate AI perpetuates, including labor exploitation and extractive, nonconsensual data collection.[10] Furthermore, ChatGPT is anthropomorphized by design, which contributes to this mystification. For instance, when a user of the tool enters a prompt, three little dots appear as if ChatGPT is "thinking," similar to a text message exchange between two friends. The tool is often deferential, apologizing when a user writes that there has been a mistake (however accurately or inaccurately). This use of anthropomorphism is deceptive, reinforcing the idea that the user is interacting with something capable of sentience or accountability, despite ChatGPT's programmed disclaimers.

While large language models (LLMs) like ChatGPT are often celebrated as being largely self-training, OpenAI researchers trained ChatGPT. They supervised and improved the model's performance using labeled data and reinforcement learning techniques. These techniques involve humans designing a system of "rewards" for the model based on desired prompt responses that ChatGPT is then optimized for.[11] Additionally, when students are asked to engage with ChatGPT, they are performing uncompensated labor for OpenAI by helping to further train the model. When they create OpenAI accounts to use ChatGPT, they provide data to the company that can be sold or shared with third parties, not to mention made vulnerable to glitches, bugs, and leaks. It's also the case that outsourced workers in Kenya were paid less than two dollars an hour to tag obscene material for removal, helping to train a detection algorithm to make its outputs less toxic.[12]

Many of the harms we might associate with ChatGPT are not inaugurated by AI. They are harms endemic to living in a society long predicated on capitalist exploitation and colonial extraction. Furthermore, as abolitionist and critical technology researcher J. Khadijah Abdurahman points out, the so-called magic of many AI tools is often predicated on evading labor protections in the academy, circumventing the hiring of student researchers by using clickworker services like Amazon Mechanical Turk and web scraping to gather and annotate data for model training.[13] Most recently, political science researchers released a non-peer-reviewed article in 2023 based on an analysis of the use of ChatGPT to classify tweets in English to perform a limited number of tasks, claiming that ChatGPT can label text more accurately than human annotation workers on Amazon Mechanical Turk. This claim erases the work of data annotators who help ensure that ChatGPT's results make sense, who are largely undertrained, undercompensated, and restricted in their ability to contribute to the development of AI tools.[14]

What might critical digital literacy education for students about ChatGPT look like in practice? Perhaps we can imagine scenarios where the use of a tool like ChatGPT can be pedagogically valuable if done using the methods and framing techniques offered by critical intellectual traditions, and if students are given the option to opt into using these tools. For instance, what might be revealed if students asked ChatGPT about its carbon footprint or to provide ten reasons not to use ChatGPT? Would ChatGPT name the ways it can be used to optimize the spread of disinformation, or how it can perpetuate racism, sexism, and Islamophobia in its responses? It's also important to remember that algorithmic harms will not weigh equally on all students. For students who already encounter systemic injustice in their daily lives, toxic algorithmic outputs that perpetuate harmful stereotypes and prejudices compound these experiences. Additionally, despite efforts to fix inaccuracies and biases in ChatGPT's responses, these harms persist. As just one example, when ChatGPT was asked to speak in the cadence of Angie Thomas, the author of *The Hate U Give*, which features an African American protagonist, ChatGPT added "yo" in front of random sentences.[15] It's also important to refuse narratives that frame AI as inevitable and minimize the power of students and educators to shape whether and how we engage with digital technology in the classroom. If educators bring ChatGPT into their classrooms, it should not be because they believe they cannot choose otherwise but because they think they can effectively leverage the tool to talk about AI's ethical implications. Sometimes, the most ethical choice is refusal.[16]

There are ways to engage students in developing students' critical digital literacy skills using ChatGPT as a case study but without having them actually experiment with the tool directly. For instance, instructional designer Autumm Caines recommends having students collectively annotate OpenAI's privacy policy and terms of service and reflect on what they notice about how student data is being used and the degree to which they consider the company transparent about its data practices.[17] Alternatively or additionally, we might imagine students conducting textual analysis of headlines surrounding ChatGPT's release and the ways they reflect certain values or assumptions about the tool or offer critical entry points for additional research. We might ask students to close read examples of ChatGPT conversations that already exist online rather than generating their own and to develop their ideas for critical public oversight, given the potential of LLMs to reproduce dominant ways of thinking about the world and denigrate the experiences and cultural contributions of people from marginalized groups.[18] We might co-facilitate discussions with students about Big Tech's hold on AI ethics research and, perhaps most relevant to conversations about ChatGPT, Google's firing of its AI ethics colead Timnit Gebru for refusing to retract her name from a paper that highlighted the social, environmental, and financial costs of LLMs.[19] In the case of ChatGPT, we might invite students to consider the large amounts of computing power, electricity, and cooling for data centers that LLMs require for model training and development. We might teach students to consider whether proposals to democratize privately funded AI tools through their dissemination in higher education is a way of allowing tech giants to continue to exercise control over the direction of AI research and perpetuate exploitative labor practices. Of course, none of these approaches are mutually exclusive, but most importantly, in a world where students are increasingly socialized to meet the demands of a hyper-extractive, surveillance-based digital economy that disproportionately harms marginalized people, we can empower one another to imagine otherwise.

Critical digital literacy education directed toward the academy can also be an occasion for students to generate counternarratives about the technology shaping their lives in solidarity with others. For instance, the teach-in #AgainstSurveillance offered an opportunity to support the legal defense fund of Ian Linkletter in the wake of Proctorio's lawsuit against him while learning more about the relationship between exam proctoring software, injustice, and organizing for social change. Similarly, the #AnnotatedEdTech event at the 2022 Civics of Technology Conference guided participants in using the social annotation tool Hypothesis to collectively annotate edtech websites to counter misleading and harmful company claims.

Contributors to Civics of Technology, a cross-campus initiative aimed at supporting students with tools and frameworks for critical technology education, has also designed three approaches for helping students interrogate the ethics of education technologies in the classroom. These approaches include guidance for conducting ethical audits of technology and discriminatory design audits. They also offer guiding questions that stress evaluating whom a given technology harms and benefits, assessing unintended or unexpected impacts, and accounting for how the technology shapes human behavior and society. These exercises help students and faculty think together about whether to use a given technology as is, modify its settings or uses, or not use it at all.[20] It's also valuable to stress to students how the idea of "unintended consequences" can be leveraged by technologists and designers to deny accountability or responsibility. This term can also elide the work happening outside of technical fields that does, in fact, have frameworks for anticipating and acknowledging the possible social, environmental, and political impacts of a given innovation.[21]

In addition to supporting students in developing critical digital literacy skills that they can apply through, and to, the university, others have proposed more robust participatory design when it comes to the development of digital tools at universities. As education scholar Neil Selwyn points out in the context of education technology, these tools are often something that is "'done to' students, educators, the less privileged and less affluent, rather than 'done by' them."[22] Selwyn and others have argued that students, teachers, and other grassroots actors should have decision-making power over how digital technology is transformed to be more equitable. This could include teachers having more involvement in the design and development of these technologies or generating their own open-source communities with students to exercise responsibility and control over the technologies shaping life in the academy. The City University of New York (CUNY) Academic Commons provides one model of what open-source higher education communities could look like if they include faculty, students, and staff in the work of developing counter-interventions to proprietary, for-profit pedagogical infrastructure.[23] The aims and objectives of open-source tools might look very different from the examples of technologies that enact punitive surveillance and austerity politics covered in this book, perhaps turning the object of inquiry for these tools from student and faculty behavior and toward redressing long-standing forms of institutionalized injustice in higher education.

Some have taken creative design approaches to help academic communities subvert the logics underpinning smart university initiatives along these lines. For instance, sociologist Juan Pardo-Guerra created rateyourcampusadmin.com to call attention to the existing power asymmetries between how administrators are typically evaluated and the heightened levels of scrutiny and metrification that faculty often face.[24] This website flips dominant power dynamics in the academy on their head, turning administrators into the primary object of scrutiny and inviting participation from faculty, staff, and students in naming administrative abuses of power. Pardo-Guerra's website is reminiscent of data scientist Brian Clifton and artists Sam Lavigne and Francis Tseng's "White Collar Crime Risk Zones," a satirical tool that uses machine learning to predict where financial crimes are likely to occur in the United States as opposed to the "street crimes" predictive policing typically targets.[25] Perhaps unsurprisingly, the Financial District of Lower Manhattan is considered a hotspot. The piece raises the question of what criminalized behaviors do and don't fall within the typical parameters of predictive policing, and what this can tell us about the built-in assumptions, and racial and economic injustices, of predictive policing tools.

It would also be beneficial to have public policy that focuses not on education as a vehicle for developing, disseminating, or keeping pace globally with AI innovation (of which there is already significant attention) but on whether and how to implement technologies like AI into higher education. Although some countries have AI policy visions that include discussions of AI for education, these discussions are generally superficial and overwhelmingly positive, failing to explore issues of informed consent, discrimination, accountability, privatization, loss of student and faculty autonomy over their learning and working conditions, and other harms.[26] However, without a fundamental reimagining of higher education, it is unlikely that the increased introduction of AI tools into universities will help offset privatization, faculty precarity, institutionalized racism, and rising student debt.

Perhaps, then, the best solution for redressing the harms of smart university initiatives is a coalition-based movement against the encroachment of the private tech industry on the university that simultaneously acknowledges higher education's deeply rooted complicity in reproducing inequality.

Such an effort would include a refusal to participate in the production and use of digital tools and practices that harm marginalized people both within and beyond the academy, and a mobilization against the institutional politics of austerity. The final chapter of this book explores how such a coalition might take shape. It builds on existing university struggles to offer alternative visions and practices for countering the smart university.

Conclusion

Against the Smart University

This book has focused primarily on describing the range of ways that students are impacted by the increasing "smartness" of their universities. However, it's also the case that faculty's working conditions are being transformed by similar tools and imperatives. The present reshaping of the university around the production of digital data has gone hand in hand with the quantification of faculty labor, including citation counts, impact factors, and performance metrics, alongside the erosion of academic freedom protections like tenure.[1] Tools like learning management systems have also been instrumental in turning teaching and learning into discrete, interchangeable, and transferable tasks that lend themselves to shifts toward contingent labor in the academy.[2] Rather than allowing tools like plagiarism detection software to set up antagonisms between faculty and students to the benefit of the private tech industry, faculty and students can find common cause in pushing back against the technocratic management of their institutions, which imposes austerity and impedes equitable learning and working conditions.

For instance, AspirEDU's Instructor Insight is a software solution that integrates with a given university's learning management system to provide administrators with metrics on instructor engagement and performance, including "frequency of course login, timeliness of grading and grade distribution."[3] According to AspirEDU's promotional website, this tool allows for instructors to be compared to one another, as well as for administrators to see and define "normal" instructor actions and activity. Although the tool is promoted as labor saving for administrators, it creates heightened levels of scrutiny for faculty and imposes quantitative performance measures that fail to account for the range of pedagogical styles that can support student learning. Similarly, Creatrix Campus's Faculty Management System promises to be labor-saving for both faculty and administrators, automating "every step of the faculty lifecycle from on-boarding, recruitment, to workload management and performance tracking, for greater institutional insights."[4] In a promotional blog post for the system, content developer Mary Clotilda Suvin writes that "the faculty performance tree assigns weights to parameters such as teaching, papers, journals published, and other information. Based on the input, the system computes an overall score and provides one-click faculty activity reporting, which assists in determining overall faculty productivity, including any setbacks."[5] Rather than reducing faculty workload in a context of stagnating wages and intensifying workplace expectations, these tools promise to help faculty use their time "well" based on parameters they do not exercise control over in order to maximize productivity.

These systems do not merely record faculty behavior but, rather, help construct the very notions by which behavior is documented and assessed, including how worth is assigned to various aspects of faculty labor. In the case of Creatrix Campus, its suite of tools includes RFID-based attendance tracking and other biometric devices to monitor how faculty use time and space. These metrics do not account for how faculty are differently situated based on race, class, ability, and gender, including how varied life circumstances and structural inequities shape productivity, promotion, salary, and citation gaps, or the fact that labor in the academy is often shared and multidimensional in practice.[6] Additionally, education technology scholars Michael Kwet and Paul Prinsloo predict a coming wave of emotion-tracking-based tools that are likely to further transform faculty working conditions in ways that intensify surveillance, including for teacher analytics and monitoring student engagement.[7]

What might it look like for students and faculty to build solidarity around resisting smart university initiatives? There is historical precedent for organizing around the production of harmful technology within the academy, particularly during the 1960s in the United States. For instance, in 1969, students, faculty, and staff organized lab walkouts at Stanford University to join a university-wide teach-in in protest of military research contracts during the Vietnam War. This period also coincided with student-led demands for curricular reform that resulted in the institutionalization of ethnic studies and women's and gender studies, albeit in ways that ultimately helped to manage and discipline the social upheavals of the period.[8] Such a case provides important lessons for how students might demand the robust integration of technology ethics education today in ways that support rather than undermine struggles for a more just technologically mediated world. For instance, rather than strictly teaching formalized rules of professional conduct and ethics as problems of individual decision making, students could be trained to consider how institutions incentivize, order, and pattern the behavior of technologists in ways that reinforce systems of oppression, which can be challenged through both individual and collective action.[9]

The late 1960s and early '70s also saw sustained organizing within computing to oppose military work, to demand greater privacy protections, and to address issues of workforce discrimination. Computer People for Peace, started in 1968 in New York, worked to raise consciousness through pamphlets, booklets, activism at major computing conferences, and raising bail for Sundiata Acoli, a Black Panther. The 1970s also saw the rise of the Polaroid Revolutionary Workers Movement, which called for an international boycott of the company given its sale of photography equipment used for enforcing the apartheid regime in South Africa.[10] These efforts reverberate into present-day tech worker organizing, including the 2018 Google walkout against sexual harassment and open letter protesting Project Maven, an AI-driven military program for interpreting video images that can be used to refine drone strikes. That same year, using the banner of #TechWontBuildIt, workers at companies including Microsoft, Amazon, IBM, and Salesforce collectively pressured Big Tech companies into ending contracts for providing information systems, data analytics, and AI to ICE and law enforcement. Microsoft and Amazon workers organized #NoTechForICE protests with immigrant rights organizations at Microsoft stores in several US cities.[11] More recently, tech workers and community activists held a #NoTechforApartheid protest at the 2023 Google Cloud conference opposing Project Nimbus, a cloud computing contract between Google, Amazon Web Services, and the Israeli military.[12] Organizing in both academia and tech workplaces is a way of checking industry power and demanding public universities that are equipped to be responsive to, rather than complicit in, technology-aided structural violence.

University researchers, for their part, could engage in technological development more reflectively, taking as a starting point the conditions, experiences, and participation of those who are most marginalized. A design justice approach would require researchers to account for how power is embedded within the research process and the potential social impacts of a given technology in order to develop community-led and nonexploitative tools.[13] It would be a research ethics that goes beyond questions of liability and Institutional Review Board approval and instead foregrounds the ways that technologies might contribute to the epistemological and material violences that sustain systems of oppression. Faculty and students can also work together to create more opportunities for technology research in solidarity with social movements outside the academy. Grassroots organizers have been successful in countering repressive technologies in their communities and have sometimes received support, funding, and labor from academics and university departments. One such example is the Our Data Bodies project, which investigates the relationship between data and socioeconomic disparity in Charlotte, Detroit, and Los Angeles.[14]

Organizing efforts against smart university initiatives are already taking shape across the country, some examples of which have been discussed in this book. Students are organizing against the remote proctoring software Proctorio, given its criminalizing logics, inaccuracies, and punitive consequences for disabled students and students of color. At UCLA, students protested university faculty participation in predictive policing technology research. Mathematics faculty from several institutions subsequently refused to collaborate with police departments in the wake of nationwide protests against anti-Black racism and police violence. As private software systems for monitoring faculty performance increasingly proliferate in US universities, faculty and students might hone their strategic efforts by studying examples such as the successful pushback against the Raising the Bar initiative at Newcastle University in England. By marshaling the support of students and parents and using emails signed by multiple scholars, flyers, and social media outreach across national boundaries, activists took collective action to fight the use of outcomes-based performance management technologies to evaluate and punish university employees.[15]

Given the relationship between smart university initiatives, austerity, and institutionalized racism in the United States, it is also crucial to pair pushback against repressive technologies used or produced in higher education with movements for the full cancellation of federal and private student loans for college. Over the past thirty years, public funding for higher education has precipitously declined while the cost of attending a four-year college has skyrocketed by almost 200 percent between 1997 and 2017, with student debt currently reaching over $1.6 trillion.[16] Students and allies are using social media among a suite of organizing tactics to demand tuition-free college and the cancellation of student debt. Black community and civic organizations have been leading this effort and are demanding greater accountability for higher education partnerships with private lenders.[17] Additionally, decentralized networks of debt resistance like the Strike Debt movement have resulted in organized efforts including the Rolling Jubilee and the Debt Collective's purchase of student loan debt, a student debt strike, and calls for the full cancellation of federal student debt through executive order.[18] Both the Debt Collective and the Student Loan Debt Center have argued that a president does have the legal authority to broadly cancel student debt, despite the June 2023 Supreme Court ruling that struck down the Biden administration's student debt relief program on the heels of the court overturning affirmative action.[19]

It's important to stress that all workers in the academy—faculty, staff, students—are impacted by the casualization, outsourcing, speedups, and finance structures dominating public higher education in the United States. Solidarity between these groups gave rise to the first Higher Ed Labor Summit in 2021, which was also attended by the Debt Collective and Scholars for a New Deal for Higher Education. Together, they developed a national labor organization called Higher Ed Labor United, whose goal is to make public colleges and universities truly public goods.[20] Davarian Baldwin poignantly stresses that faculty and staff must firmly situate their bread-and-butter concerns within the context of "broader university-related struggles around housing displacement, student debt, policing, contingent labor, reparations, food insecurity, payment in lieu of taxes, equitable STEM research, endowment equity, and donor allocations," and to which we might add digitization initiatives that support and sustain the harmful exercise of university power over students, faculty, staff, and communities.[21] Such initiatives can and should be sites of union bargaining given their implications for the quality and security of academic working conditions.

This book has argued that campus modernization efforts often intensify the politics of austerity in higher education, and thus university efforts to monitor and automate learning and labor conditions cannot go unchecked. The case of West Virginia University is particularly instructive here, which recently made wide-ranging cuts to humanities academic programs and faculty positions, in partnership with consulting firm rpk Group, after rapid withdrawals of state funding from the school. These cuts occurred after a decade of extensive new campus construction initiatives funded by debt and private-public partnerships, as well as recent commitments to overhaul and modernize using data-intensive technologies across all levels of the university.[22] As universities increasingly incorporate digital technologies to automate academic labor, surveil campus communities, and embed austerity logics into multiple facets of university governance, struggling for the

Conclusion: Against the Smart University

future of higher education includes demanding democratic control over whether and how digital tools get used.

Ultimately, this book has sought to contribute to visions of how we might reckon with universities' longstanding complicity in structural injustice and the role that digital technologies are playing today in reproducing oppression within and through the academy. Drawing from Fred Moten and Stefano Harney's *The Undercommons: Fugitive Planning and Black Study*, Abigail Boggs, Eli Meyerhoff, Nick Mitchell, and Zach Schwartz-Weinstein approach the university as a site of abolitionist struggle and a space for imagining a new society.[23] As this book has demonstrated, smart university initiatives often perpetuate the logics and tools underpinning racial injustice, from constructions of "at-risk" student populations to the design and dissemination of repressive, anti-Black technologies. Pushback against the smart university, including the development of technologies that discipline, surveil, and punish, is a struggle that can and should be linked to existing community-driven abolitionist organizing. To push back against the smart university also means countering the underfunding and institutional marginalization of critical approaches to technology, and demanding research in solidarity with efforts outside the academy for radical technological reform until the academy as we know it today is fundamentally transformed. A different university, one that empowers people to carry out the social changes necessary for a more just world, is possible.

Notes

Introduction

1. Nancy L. Zimpher, "Building a Smarter University: Big Data, Innovation, and Ingenuity," in *Building a Smarter University: Big Data, Innovation, and Analytics*, ed. Jason E. Lane (Albany: State University of New York Press, 2014), xi–xv.

2. Pauline Hope Cheong and Pratik Nyaupane, "Smart Campus Communication, Internet of Things, and Data Governance: Understanding Student Tensions and Imaginaries," *Big Data & Society* 9, no. 1 (2022): 2, https://doi.org/10.1177/20539517221092656.

3. See, for example, Martin Jones, "What Is a Smart Campus and the Benefits to College Students and Faculty," *Cox Business*, accessed May 21, 2023, https://www.coxblue.com/what-is-a-smart-campus-and-the-benefits-to-college-students-and-faculty/; Vladimir L. Uskov, Jeffrey P. Bakken, Robert James Howlett, and Lakhmi C. Jain, eds., *Smart Universities: Concepts, Systems, and Technologies* (Cham, Switzerland: Springer International, 2018); and Jason E. Lane, ed., *Building a Smarter University: Big Data, Innovation, and Analytics* (Albany: State University of New York Press, 2014).

4. Sheila Jasanoff and Sang-Hyun Kim, *Dreamscapes of Modernity: Sociotechnical Imaginaries and the Fabrication of Power* (Chicago: University of Chicago Press, 2015), 153.

5. Orit Halpern and Robert Mitchell, *The Smartness Mandate* (Cambridge, MA: MIT Press, 2022).

6. Meredith Broussard, *Artificial Unintelligence: How Computers Misunderstand the World* (Cambridge, MA: MIT Press, 2018).

7. Nasro Min-Allah and Saleh Alrashed, "Smart Campus—a Sketch," *Sustainable Cities and Society* 59 (2020): 3, https://doi.org/10.1016/j.scs.2020.102231.

8. Sheila Slaughter and Gary Rhoades, "The Neo-liberal University," *New Labor Forum*, no. 6 (2000): 73–79, https://www.jstor.org/stable/40342886.

9. Deborah Lupton, *Digital Sociology* (London: Routledge, 2015).

10. Jones, "What Is a Smart Campus and the Benefits to College Students and Faculty."

11. Ben Williamson, "The Hidden Architecture of Higher Education: Building a Big Data Infrastructure for the 'Smarter University,'" *International Journal of Higher Education* 15, no. 12 (2018): 1–26, https://doi.org/10.1186/s41239-018-0094-1; Neil Selwyn, *Digital Technology and the Contemporary University: Degrees of Digitization* (London: Routledge, 2014); Audrey Watters, "Ed-Tech and the Californian Ideology," *Hack Education*, last modified May 17, 2015, http://hackeducation.com/2015/05/17/ed-tech-ideology.

12. Steven Lubar, "'Do Not Fold, Spindle or Mutilate': A Cultural History of the Punch Card," *Journal of American Culture* 15, no. 4 (1992): 46.

13. Mario Savio, "Sit-In Address on the Steps of Sproul Hall" (speech delivered at the University of California, Berkeley, December 2, 1964).

14. Lubar, "'Do Not Fold, Spindle, or Mutilate.'"

15. Elana Zeide, "Student Privacy Principles for the Age of Big Data: Moving Beyond FERPA and FIPPS," *Drexel Law Review* 8, no. 339 (2016): 354. This historical overview of FERPA is significantly indebted to Zeide's scholarship.

16. Diane Divoky, "Cumulative Records: Assault on Privacy," *Learning* 2, no. 18 (1973): 18–21.

17. Karen J. Stone and Edward N. Stoner II, "Revisiting the Purpose and Effect of FERPA," Stetson University College of Law 23rd Annual National Conference on Law and Higher Education (February 2002): 2, http://www.stetson.edu/law/academics/highered/home/media/2002/Revisiting_the_Purpose_of_FERPA.pdf.

18. Virginia Eubanks, "Want to Predict the Future of Surveillance? Ask Poor Communities," *American Prospect*, January 15, 2014, https://prospect.org/power/want-predict-future-surveillance-ask-poor-communities./.

19. Ariana Torbin, "HUD Sues Facebook over Housing Discrimination and Says the Company's Algorithms Have Made the Problem Worse," *ProPublica*, last modified March 18, 2019, https://www.propublica.org/article/hud-sues-facebook-housing-discrimination-advertising-algorithms.

20. Stone, "Revisiting the Purpose and Effect of FERPA," 10.

21. Zeide, "Student Privacy Principles for the Age of Big Data," 368.

22. For an overview of the field of science and technology studies, see Ulrike Felt, Rayvon Fouché, Clark A. Miller, and Laurel Smith-Doerr, eds., *The Handbook of Science and Technology Studies*, 4th ed. (Cambridge, MA: MIT Press, 2016).

23. Pierre Lascoumes and Patrick Le Galès, "Understanding Public Policy through Its Instruments—from the Nature of Instruments to the Sociology of Public Policy Instrumentation," *Governance: An International Journal of Policy, Administration, and Institutions* 20, no. 1 (2007): 1–21.

24. Jeffery Kahn, "Ronald Reagan Launched Political Career Using the Berkeley Campus as a Target," *UC Berkeley News*, last modified June 8, 2004, https://www.berkeley.edu/news/media/releases/2004/06/08_reagan.shtml.

25. Tressie McMillan Cottom, *Lower Ed: The Troubling Rise of For-Profit Colleges in the New Economy* (New York: New Press, 2017), 134.

26. Jay W. Goff and Christopher M. Shaffer, "Big Data's Impact on College Admission Practices and Recruitment Strategies," in *Building a Smarter University: Big Data, Innovation, and Analytics*, ed. Jason Lane (Albany: State University of New York Press, 2014), 108.

27. Peter Miller and Nikolas S. Rose, *Governing the Present: Administering Economic, Social and Personal Life* (Cambridge: Polity, 2008), 216.

28. The term "data double" comes from Kevin D. Haggerty and Richard V. Ericson, "The Surveillant Assemblage," *British Journal of Sociology* 51, no. 4 (2000): 605–622.

29. Advancing Minorities' Interest in Engineering, "HBCUs, Industry Partners Launch Initiative to Diversity and Strengthen Cybersecurity Workforce," Abbott (press release), accessed October 1, 2023, https://abbott.mediaroom.com/2022-06-02-HBCUs,-Industry-Partners-Launch-Initiative-to-Diversify-and-Strengthen-Cybersecurity-Workforce.

30. Daysi Diaz-Strong, Christina Gómez, María E. Luna-Duarte, Erica R. Meiners, and Luvia Valentin, "Organizing Tensions—from the Prison to the Military-Industrial-Complex," *Social Justice* 36, no. 2 (2009/2010): 74, https://www.jstor.org/stable/29768538.

31. Amy Zegart and Michael Morell, "Spies, Lies, and Algorithms: Why U.S. Intelligence Agencies Must Adapt or Fail," *Foreign Affairs*, last modified May 2019, https://www.foreignaffairs.com/articles/2019-04-16/spies-lies-and-algorithms.

32. Abigail Boggs and Nick Mitchell, "Critical University Studies and the Crisis Consensus," *Feminist Studies* 44, no. 2 (2018): 434, https://doi.org/10.15767/feministstudies.44.2.0432.

33. Christopher Newfield, *The Great Mistake: How We Wrecked Public Universities and How We Can Fix Them* (Baltimore: John Hopkins University Press, 2016), 23.

34. Newfield, *The Great Mistake*, 11.

35. Newfield, *The Great Mistake*, 17.

36. Newfield, *The Great Mistake*, 24.

Chapter 1. The "Smart" University

1. Deloitte, "Smart Campus: The Next-Generation Connected Campus," last modified 2019, https://www2.deloitte.com/content/dam/Deloitte/us/Documents/strategy/the-next-generation-connected-campus-deloitte.pdf.

2. Quoted in Julie Rubley, "Achieving the Smart Campus of Tomorrow Today," *Educause Review*, last modified November 25, 2019, https://er.educause.edu/blogs/sponsored/2019/11/achieving-the-smart-campus-of-tomorrow-today.

3. Nancy L. Zimpher, "Building a Smarter University: Big Data, Innovation, and Ingenuity," in *Building a Smarter University: Big Data, Innovation, and Analytics*, ed. Jason E. Lane (Albany: State University of New York Press, 2014), xi–xv.

4. Zhao Yang Dong, Yuchen Zhang, Christine Yip, Sharon Swift, and Kim Beswick, "Smart Campus: Definition, Framework, Technologies, and Services," *Institution of Engineering and Technology* (2020), pp. 1–12, https://doi.org/10.1049/iet-smc.2019.0072.

5. Ben Williamson, "Psychodata: Disassembling the Psychological, Economic, and Statistical Infrastructure of 'Socio-emotional Learning,'" Journal of Education Policy 36, no. 1 (2021): 135. https://doi.org/10.1080/02680939.2019.1672895.

6. Martin Jones, "What Is a Smart Campus and the Benefits to College Students and Faculty," *Cox Business*, accessed May 21, 2023, https://www.coxblue.com/what-is-a-smart-campus-and-the-benefits-to-college-students-and-faculty/.

7. Deloitte, "Smart Campus."

8. Nasro Min-Allah and Saleh Alrashed, "Smart Campus—a Sketch," *Sustainable Cities and Society* 59 (2020): 1–15, https://doi.org/10.1016/j.scs.2020.102231.

9. Jason Lane and B. Alex Finsel, "Fostering Smarter Colleges and Universities: Data, Big Data, and Analytics," in *Building a Smarter University: Big Data, Innovation, and Analytics*, ed. Jason Lane (Albany: State University of New York Press, 2014), 3–25.

10. David Lyon, *The Culture of Surveillance: Watching as a Way of Life* (New York: John Wiley & Sons, 2018), 6.

11. Rita Raley, "Dataveillance and Counterveillance," in *Raw Data Is an Oxymoron*, ed. Lisa Gitelman (Cambridge, MA: MIT Press, 2013), 124.

12. Armand Mattelart, *The Information Society* (London: Sage, 2003), 36.

13. Oscar H. Gandy, *The Panoptic Sort: A Political Economy of Personal Information* (Boulder, CO: Westview, 1993).

14. Nelson Lichtenstein, *The Retail Revolution: How Wal-Mart Created a Brave New World of Business* (New York: Metropolitan, 2009), 57.

15. Dan Schiller, *Digital Depression: Information Technology and Economic Crisis* (Champaign: University of Illinois Press, 2014), 129.

16. Stephanie Ricker Schulte, "Personalization," in *Digital Keywords* (Princeton, NJ: Princeton University Press, 2016), 249.

17. Thomas H. Davenport and John C. Beck, *The Attention Economy: Understanding the New Currency of Business* (Cambridge, MA: Harvard Business Review Press, 2002).

18. Mark Andrejevic, "Estrangement 2.0," *World Picture* 6 (2011): 1.

19. Theodor Adorno and Max Horkheimer, "The Culture Industry: Enlightenment as Mass Deception," *Dialectic of Enlightenment* (Stanford, CA: Stanford University Press, 2022), 94–136.

20. Schulte, *Digital Keywords*, 242.

21. Simone Browne, *Dark Matters: On the Surveillance of Blackness* (Durham, NC: Duke University Press, 2015), 17.

22. Meredith Broussard, *Artificial Unintelligence: How Computers Misunderstand the World* (Cambridge, MA: MIT Press, 2018).

23. Hal Varian, "Differential Pricing and Efficiency," *First Monday* 1, no. 2 (1996), http://firstmonday.org/ojs/index.php/fm/article/view/473/394.

24. Chris Gilliard, "Pedagogy and the Logic of Platforms," in *Open at the Margins: Critical Perspectives on Open Education*, ed. Maja Bali, Catherine Cronin, Laura Czerniewicz, Robin DeRosa, and Rajiv Jhangiani (Montreal, Quebec: Rebus Community, 2020), 116.

25. Gilliard, "Pedagogy and the Logic of Platforms."

26. Latanya Sweeney, "Discrimination in Online Ad Delivery," *Communications of the ACM* 56, no. 5 (2013): 44–54, https://doi.org/10.1145/2447976.2447990.

27. Astra Taylor and Jathan Sadowski, "How Companies Turn Your Facebook Activity into a Credit Score," *Nation*, May 27, 2015, https://www.thenation.com/article/archive/how-companies-turn-your-facebook-activity-credit-score/.

28. Jennifer Wriggins, "The Color of Property and Auto Insurance: Time for Change," *Florida State University Law Review* 49, no. 203 (2022): 203–256.

29. Tarleton Gillespie, "The Relevance of Algorithms," *Media Technologies*, ed. Tarleton Gillespie, Pablo Boczkowshi, and Kirsten Foot (Cambridge, MA: MIT Press, 2012), 171.

30. Bill Gates, *The Road Ahead* (London: Penguin, 1996), 180.

31. Bill Gates, "Prepared Remarks at the National Conference of State Legislatures," *Bill & Melinda Gates Foundation*, accessed October 9, 2023, https://www.gatesfoundation.org/ideas/speeches/2009/07/bill-gates-national-conference-of-state-legislatures-ncsl.

32. Cass Sunstein, "Nudging: A Very Short Guide," *Journal of Consumer Policy* 37 (2014): 583–588, https://doi.org/10.1007/s10603-014-9273-1.

33. Manuela Ekowo and Iris Palmer, "The Promise and Peril of Predictive Analytics in Higher Education: A Landscape Analysis," *New America*, October 24, 2016, https://www.newamerica.org/education-policy/policy-papers/promise-and-peril-predictive-analytics-higher-education/.

34. Neil Selwyn, *Education and Technology: Key Issues and Debates*, 2nd ed. (New York: Bloomsbury, 2016), 102–134.

35. Audrey Watters, *Teaching Machines: The History of Personalized Learning* (Cambridge, MA: MIT Press, 2021).

36. Watters, *Teaching Machines*, 70.

37. Watters, *Teaching Machines*, 160.

38. Watters, *Teaching Machines*, 96.

39. Annalisa Quinn, "Book News: Amazon Wants to Ship Products before You Even Buy Them," *The Two-Way: Breaking News from NPR*, January 20, 2014, http://www.npr.org/sections/thetwo-way/2014/01/20/264187990/book-news-amazon-wants-to-ship-products-before-you-even-buy-them.

40. Dong et al., "Smart Campus."

41. Min-Allah and Alrashed, "Smart Campus."

42. Collabco, "From Digital Student to Smart Citizen," white paper, accessed October 10, 2023, https://myday.collabco.com/wp-content/uploads/from-digital-student-to-smart-citizen-whitepaper.pdf.

43. Vito Albino, Umberto Berardi, and Rosa Maria Dangelico, "Smart Cities: Definitions, dimensions, performance, and initiatives," *Journal of Urban Technology* 22, no. 1 (2015): 3–21. https://doi.org/10.1080/10630732.2014.942092.

44. Simone Tulumello and Fabio Iapaolo, "Policing the Future, Disrupting Urban Policy Today: Predictive Policing, Smart City, and Urban Policy in Memphis (TN)," *Urban Geography* 43, no. 3 (2021): 451, https://doi.org/10.1080/02723638.2021.1887634.

45. Christopher Gaffney and Cerianne Robertson, "Smarter Than Smart: Rio de Janeiro's Flawed Emergence as a Smart City," *Journal of Urban Technology* 25, no. 3 (2016): 60, https://doi.org/10.1080/10630732.2015.1102423.

46. Monique Mann, Peta Mitchell, Marcus Foth, and Irina Anastasiu, "#BlockSidewalk to Barcelona: Technological Sovereignty and the Social License to Operate Smart Cities," *Journal of the Association for Information Science and Technology* 71, no. 9 (2020): 1107, https://doi.org/10.1002/asi.24387.

47. Adam Greenfield, *Against the Smart City* (Berlin: Do Projects, 2013), 115.

48. Deloitte, "Smart Campus," 2.

49. Deloitte, "Smart Campus," 8.

50. Zofia Niemtus, "Are University Campuses Turning into Mini Smart Cities?" *Guardian*, February 22, 2019, https://www.theguardian.com/education/2019/feb/22/are-university-campuses-turning-into-mini-smart-cities.

51. Tom Dart, "University of Texas: Eco-conscious Campus and Major Fracking Landlord," *Guardian*, October 10, 2018, https://www.theguardian.com/us-news/2018/oct/10/university-of-texas-eco-conscious-campus-and-major-fracking-landlord.

52. Stephanie Weagle, "From Safe Campus to Smart Campus," *Campus Security Today*, April 1, 2019, https://campuslifesecurity.com/Articles/2019/04/01/From-Safe-Campus-to-Smart-Campus.aspx?Page=1.

53. See, for instance, Tulumello and Iapaolo, "Policing the Future"; and Jathan Sadowski, *Too Smart: How Digital Capitalism Is Extracting Data, Controlling Our Lives, and Taking Over the World* (Cambridge, MA: MIT Press, 2020).

54. Tulumello and Iapaolo, "Policing the Future"; Rashida Richardson, Jason Schultz, and Kate Crawford, "Dirty Data, Bad Predictions: How Civil Rights Violations Impact Police Data, Predictive Policing Systems, and Justice," *New York University Law Review Online* 94, no. (2019), https://papers.ssrn.com/sol3/papers.cfm?abstract_id=3333423#.

55. Weagle, "From Safe Campus to Smart Campus."

56. Tulumello and Iapaolo, "Policing the Future," 4.

57. Siemens, "Attractive Higher Education Campuses," last modified August 2022, https://assets.new.siemens.com/siemens/assets/api/uuid:fb4dad9c-5c51-43aa-96ce-8d64128d1a44/smart-campus-university-ipdf-en.pdf.

58. National Student Clearinghouse Research Center, "Current Term Enrollment Estimates: Fall 2022 Expanded Edition," last modified February 2, 2023, https://nscresearchcenter.org/current-term-enrollment-estimates/.

59. Andrew J. Hawkins, "Alphabet's Sidewalk Labs Shuts Down Toronto Smart City Project," *Verge*, May 7, 2020, https://www.theverge.com/2020/5/7/21250594/alphabet-sidewalk-labs-toronto-quayside-shutting-down.

60. Dong et al., "Smart Campus."

61. Dong et al., "Smart Campus," 5.

62. Sadowski, *Too Smart*, 5.

63. Sam Biddle, "Amazon Admits Giving Ring Camera Footage to Police without a Warrant or Consent," *Intercept*, July 13, 2020, https://theintercept.com/2022/07/13/amazon-ring-camera-footage-police-ed-markey/.

64. Langdon Winner, "Do Artifacts Have Politics," *Daedalus* 109, no. 1 (Winter, 1980): 121–136; Ben Green, "Data Science as Political Action," *Journal of Social Computing* 2, no. 3 (September 2021): 246–265, https://doi.org/10.23919/JSC.2021.0029.

65. Lisa Gitelman and Virginia Jackson, introduction to *Raw Data Is an Oxymoron*, ed. Lisa Gitelman (Cambridge, MA: MIT Press, 2013), 6.

66. Madisson Whitman, "We Called That a Behavior: The Making of Institutional Data," *Big Data & Society* 7, no. 1, (2020): 2, https://doi.org/10.1177/2053951720932200.

67. Greenfield, *Against the Smart City*, 74.

68. Andrew G. Haldane, "Ideas and Institutions—a Growth Story" (speech to the Guild Society, University of Oxford, May 23, 2018), https://www.bis.org/review/r180627e.pdf.

69. Joseph E. Aoun, *Robot Proof: Higher Education in the Age of Artificial Intelligence* (Cambridge, MA: MIT Press, 2018).

70. Michael M. Crow and William B. Dabars, *Designing the New American University* (Baltimore: Johns Hopkins University Press, 2015), 268.

71. Christopher Newfield, "What Is New about the New American University?" *Los Angeles Review of Books*, April 5, 2015, https://lareviewofbooks.org/article/new-new-american-university/.

72. Jamie Morgan, "Will We Work in Twenty-First Century Capitalism? A Critique of the Fourth Industrial Revolution Literature," *Economy and Society* 48, no. 3 (2019): 371–398, https://doi.org/10.1080/03085147.2019.1620027.

73. See Temitayo Shenkoya and Euiseok Kim, "Sustainability in Higher Education: Digital Transformation of the Fourth Industrial Revolution and Its Impact on Open Knowledge," *Sustainability* 15, no. 3 (2023): 1–16, https://doi.org/10.3390/su15032473; Sultan Refa Alotaibi, "An Integrated Framework for Smart College Based on the Fourth Industrial Revolution," *International Transaction Journal of Engineering, Management, & Applied Sciences and Technologies* 12, no. 4 (2021): 1–18, https://tuengr.com/V12/12A4R.pdf.

74. Ian Moll, "The Fourth Industrial Revolution: A New Ideology," *Triple C: Communication, Capitalism & Critique* 20, no. 1 (February, 2022): 47, https://doi.org/10.31269/triplec.v20i1.1297.

75. Aaron Benanav, *Automation and the Future of Work* (London: Verso Books, 2022).

76. Moll, "The Fourth Industrial Revolution," 54.

Chapter 2. Recruitment

1. Peter Coy, "The College Admissions Scandal Presses Our 'Unfairness' Button," *Bloomberg*, March 14, 2019, https://www.bloomberg.com/news/articles/2019-03-14/the-college-admissions-scandal-presses-our-unfairness-button.

2. Robert Lee, Tristan Ahtone, Margaret Pearce, Kalen Goodluck, Geoff McGhee, Cody Leff, Katherine Lanpher, and Taryn Salinas, "Land Grab Universities," *High Country News*, last modified 2020, https://www.landgrabu.org/.

3. Amaka Okechukwu, *To Fulfill These Rights: Political Struggle over Affirmative Action & Open Admissions* (New York: Columbia University Press, 2019), 37.

4. John Schwarz, "The Origin of Student Debt: The Danger of Educated Proles," *Intercept*, August 25, 2022, https://theintercept.com/2022/08/25/student-loans-debt-reagan/.

5. W. Carson Byrd, *Behind the Diversity Numbers: Achieving Racial Equity on Campus* (Boston: Harvard Education Press, 2021).

6. See Students for Fair Admissions v. President and Fellows of Harvard College, No. 20–1199, slip op. (U.S. Oct. 31, 2022), https://www.supremecourt.gov/opinions/22pdf/20-1199_hgdj.pdf; and Brief for the United States as Amicus Curiae Supporting Respondents, Students for Fair Admissions v. University of North Carolina et al., 600 U.S. ____ (2023) (No. 21–707), https://www.justice.gov/crt/case-document/file/1523991/download.

7. Dominique J. Baker, "Pathways to Racial Equity in Higher Education: Modeling the Antecedents of State Affirmative Action Bans," *American Educational Research Journal* 56, no. 5 (2019): 1861–1895, https://doi.org/10.3102/0002831219833918.

8. Baker, "Pathways to Racial Equity."

9. Department of Justice and Department of Education, "Questions and Answers Regarding the Supreme Court Decision in *Students for Fair Admissions, Inc. v. Harvard College and University of North Carolina*," August 14, 2023, https://www.justice.gov/d9/2023-08/post-sffa_resource_faq_final_508.pdf.

10. Sara Ahmed, *On Being Included: Racism and Diversity in Institutional Life* (Durham, NC: Duke University Press, 2012).

11. Okechukwu, *To Fulfill These Rights*, 168.

12. Byrd, *Behind the Diversity Numbers*, 6.

13. PBS Frontline, "Where Did the Test Come From? Americans Instrumental in Establishing Standardized Tests," accessed October 19, 2023, https://www.pbs.org/wgbh/pages/frontline/shows/sats/where/three.html.

14. Ezekiel J. Dixon-Román, *Inheriting Possibility: Social Reproduction and Quantification in Education* (Minneapolis: University of Minnesota Press, 2017), 142.

15. Mark DeGeurin, "The College Board Is Licensing the Personal Data of Students Taking the SAT to Colleges So They Can Reject More Students and Inflate Admissions Numbers," *Insider*, November 6, 2019, https://www.insider.com/college-board-sat-student-data-colleges-to-reject-students-admissions-2019-11.

16. Capture, "The 1-2-3 of Machine Learning to Power Your Predictive Model," accessed August 24, 2021, https://www.youtube.com/watch?v=5p5cYV5AtB0.

17. Capture, "Make Every Interaction Count," accessed May 21, 2023, https://www.capturehighered.com/.

18. Capture, "Make Every Interaction Count."

19. Ben Williamson, "Digital Education Governance: Data Visualization, Predictive Analytics, and 'Real-Time' Policy Instruments," *Journal of Education Policy* 31, no. 2 (April 2015): 123–141, https://doi.org/10.1080/02680939.2015.1035758.

20. National Student Clearinghouse Research Center, "Completing College—National by Race and Ethnicity—2017," last modified April 26, 2017, https://nscresearchcenter.org/signaturereport12-supplement-2/.

21. Capture, "Delivering the Industry's Highest-Qualified Student Inquiries," accessed May 21, 2023, https://www.capturehighered.com/wp-content/uploads/2021/09/Capture-Inquiries-Sales.pdf.

22. National Student Clearinghouse Research Center, "Completing College."

23. Marian Wang, "Public Universities Ramp Up Aid for the Wealthy, Leaving the Poor Behind," *ProPublica*, September 11, 2013, https://www.propublica.org/article/how-state-schools-ramp-up-aid-for-the-wealthy-leaving-the-poor-behind/.

24. CampusLogic, "Higher Education Financial Aid Software," accessed May 21, 2023, https://campuslogic.com/.

25. Capture, "Marketing & Recruitment Solutions," accessed June 20, 2022, https://www.capturehighered.com/solutions/.

26. Capture, "Marketing & Recruitment Solutions."

27. Capture, "The 1-2-3 of Machine Learning."

28. Capture, "The 1-2-3 of Machine Learning."

29. Ben Green, "'Fair' Risk Assessments: A Precarious Approach for Criminal Justice Reform," in *5th Workshop on Fairness, Accountability, and Transparency in Machine Learning* (2018): 1–5, https://scholar.harvard.edu/files/bgreen/files/18-fatml.pdf.

30. Julia Powles and Helen Nissenbaum, "The Seductive Diversion of 'Solving Bias' in Artificial Intelligence," *OneZero*, last modified December 7, 2019, https://onezero.medium.com/the-seductive-diversion-of-solvingbias-in-artificial-intelligence-890df5e5ef53; Roel Dobbe, Sarah Dean, Thomas Gilbert, and Nitin Kohli, "A Broader View on Bias in Automated Decision-Making: Reflecting on Epistemology and Dynamics," in *Proceedings of ICML Workshop on Fairness, Accountability, and Transparency in Machine Learning* (2018): 1–5, https://arxiv.org/pdf/1807.00553.pdf.

31. Amber Jamilla Musser, "Specimen Days: Diversity, Labor, and the University," *Feminist Formations* 27, no. 3 (2015): 1–20, https://www.jstor.org/stable/43860813.

32. Ted Thornhill, "We Want Black Students, Just Not You: How White Admissions Counselors Screen Black Prospective Students," *Sociology of Race and Ethnicity* 5, no 4 (October 2019): 456–470, https://doi.org/10.1177/2332649218792579.

33. Natasha Singer, "They Loved Your G.P.A. Then They Saw Your Tweets," *New York Times*, last modified November 9, 2013, https://www.nytimes.com/2013/11/10/business/they-loved-your-gpa-then-they-saw-your-tweets.html.

34. Brooke Erin Duffy and Ngai Keung Chan, "'You Never Really Know Who's Looking': Imagined Surveillance across Social Media Platforms," *New Media & Society*, no. 1 (2019): 119–138, https://doi.org/10.1177/1461444818791318.

35. Abigail Boggs, "On Borders and Academic Freedom: Noncitizen Students and the Limits of Rights," *AAUP Journal of Academic Freedom* 11 (2020): 6.

36. Boggs, "On Borders and Academic Freedom."

37. Boggs, "On Borders and Academic Freedom," 8.

38. Caitlin Dickerson, "'Demeaned and Humiliated': What Happened to These Iranians at US Airports," *New York Times*, last modified January 27, 2020, https://www.nytimes.com/2020/01/25/us/iran-students-deported-border.html.

39. Karen Zraick and Mihir Zaveri, "Harvard Student Says He Was Barred from US over His Friends' Social Media Posts," *New York Times*, August 27, 2019, https://www.nytimes.com/2019/08/27/us/harvard-student-ismail-ajjawi.html.

40. Collabco, "From Digital Student to Smart Citizen," white paper, accessed October 10, 2023, https://myday.collabco.com/wp-content/uploads/from-digital-student-to-smart-citizen-whitepaper.pdf.

41. Okechukwu, *To Fulfill These Rights*, 3.

42. Judith Scott-Clayton and Jing Li, "Black-White Disparity in Student Loan Debt More Than Triples after Graduation," *Brookings*, last modified October 20, 2016, https://www.brookings.edu/research/black-white-disparity-in-student-loan-debt-more-than-triples-after-graduation/.

Chapter 3. Retention

1. My use of the term "objects of inquiry" is borrowed from Craig Robertson's *The Passport in America: The History of a Document* (Oxford: Oxford University Press, 2010), which uses this term to describe the transformation of identity under twentieth-century US nation-state surveillance.

2. Jimmy Sanderson, Blair Browning, and Annelie Schmittel, "Education on the Digital Terrain: A Case Study Exploring College Athletes' Perceptions of Social-Media Training," *International Journal of Sport Communication* 8, no. 1 (2015): 103–124, https://doi.org/10.1123/IJSC.2014-0063.

3. Michael K. Park, "'Stick to Sports'? First Amendment Values and Limitations to Student-Athlete Expression," *Journalism & Mass Communication Quarterly* 99, no. 2 (June 2022): 518, https://doi.org/10.1177/10776990211018757.

4. Eddie Comeaux, "Stereotypes, Control, Hyper-surveillance, and Disposability of the NCAA Division I Black Male Athletes," *New Directions for Student Services*, no. 163 (2018): 39.

5. Comeaux, "Stereotypes, Control, Hyper-surveillance, and Disposability of the NCAA Division I Black Male Athletes."

6. Andrew McGregor, "Black Labor, White Profits, and How the NCAA Weaponized the Thirteenth Amendment," *Sport in American History*, March 1, 2018, https://ussporthistory.com/2018/03/01/black-labor-white-profits-and-how-the-ncaa-weaponized-the-thirteenth-amendment/.

7. Philip Doty, "Library Analytics as Moral Dilemmas for Academic Librarians," *Journal of Academic Librarianship* 46, no. 4, (July 2020): 1–5.

8. Rodrigo Ochigame, "The Long History of Algorithmic Fairness," *Phenomenal World*, January 30, 2020, https://www.phenomenalworld.org/analysis/long-history-algorithmic-fairness/.

9. Marion Fourcade and Kieran Healy, "Classification Situations: Life-Chances in the Neoliberal Era," *Accounting, Organizations and Society* 38, no. 8 (November 2013): 559, https://doi.org/10.1016/j.aos.2013.11.002.

10. Ochigame, "The Long History."

11. Ruha Benjamin, *Race after Technology: Abolitionist Tools for the New Jim Code* (Cambridge: Polity, 2019).

12. Manuela Ekowo and Iris Palmer, "The Promise and Peril of Predictive Analytics in Higher Education: A Landscape Analysis," *New America*, October 24, 2016, https://www.newamerica.org/education-policy/policy-papers/promise-and-peril-predictive-analytics-higher-education/, 8.

13. Ben Green and Lily Hu, "The Myth in the Methodology: Towards a Recontextualization of Fairness in Machine Learning," in *Machine Learning: The Debates Workshop, 35th International Conference on Machine Learning* (2018): 1–5, https://econcs.seas.harvard.edu/files/econcs/files/green_icml18.pdf.

14. Michael Kwet and Paul Prinsloo, "The 'Smart' Classroom: A New Frontier in the Age of the Smart University," *Teaching in Higher Education* 25, no. 4 (2020): 510–526, https://doi.org/10.1080/13562517.2020.1734922.

15. Shirin Ghaffary, "Amazon Fired Chris Smalls: Now the New Union Leader Is One of Its Biggest Problems," *Vox*, June 7, 2022, https://www.vox.com/recode/23145265/amazon-fired-chris-smalls-union-leader-alu-jeff-bezos-bernie-sanders-aoc-labor-movement-biden.

16. Ben Williamson, "The Hidden Architecture of Higher Education: Building a Big Data Infrastructure for the 'Smarter University,'" *International Journal of Higher Education* 15, no. 12 (2018): 5, https://doi.org/10.1186/s41239-018-0094-1.

17. Drew Harwell, "Colleges Are Turning Students' Phones into Surveillance Machines, Tracking the Locations of Hundreds of Thousands," *Washington Post*, December 24, 2019, https://www.washingtonpost.com/technology/2019/12/24/colleges-are-turning-students-phones-into-surveillance-machines-tracking-locations-hundreds-thousands/.

18. Olga Viberg, Mathias Hatakka, Olof Balter, and Anna Mavroudi, "The Current Landscape of Learning Analytics in Higher Education," *Computers in*

Human Behavior 89 (December 2018): 98–110, https://doi.org/10.1016/j.chb.2018.07.027.

19. Kwet and Prinsloo, "The 'Smart' Classroom."

20. Beckie Supiano, "Nudging Looked Like It Could Help Solve Key Problems in Higher Ed: Now That's Not so Clear," *Chronicle of Higher Education*, September 4, 2019, https://www.chronicle.com/article/nudging-looked-like-it-could-help-solve-key-problems-in-higher-ed-now-thats-not-so-clear/; Peter Francis, Christine Broughan, Carly Foster, and Caroline Wilson, "Thinking Critically about Learning Analytics, Student Outcomes, and Equity of Attainment," *Assessment & Evaluation in Higher Education* 45, no. 6 (December 2019): 811–821, https://doi.org/10.1080/02602938.2019.1691975.

21. Kwet and Prinsloo, "The 'Smart' Classroom," 519.

22. Sun-ha Hong, "Predictions without Futures," *History and Theory* 61, no. 3 (August 2022): 375, https://doi.org/10.1111/hith.12269.

23. Lee Gardner, "Students under Surveillance? Data-Tracking Enters a Provocative New Phase," *Chronicle of Higher Education*, October 13, 2019, https://www.chronicle.com/article/students-under-surveillance/.

24. Jill Barshay and Sasha Aslanian, "Under a Watchful Eye," *AMP Reports*, August 9, 2019, https://www.apmreports.org/episode/2019/08/06/college-data-tracking-students-graduation.

25. Adrianna Kazar and Daniel Maxey, "The Changing Faculty and Student Success: Selected Research on Connections between Non-tenure-track Faculty and Student Learning," Pulias Center for Higher Education (2012), https://files.eric.ed.gov/fulltext/ED532273.pdf.

26. Jeremy Knox, Ben Williamson, and Sian Bayne, "Machine Behaviourism: Future Visions of 'Learnification' and 'Datafication' across Human and Digital Technologies," *Learning, Media and Technology* 45, no. 1 (2020): 34, https://doi.org/10.1080/17439884.2019.1623251.

27. Evgeny Morozov, *To Save Everything Click Here: Technology, Solutionism and the Urge to Fix Problems That Don't Exist* (London: Penguin, 2013).

28. Vimal Patel, "Are Students Socially Connected? Check Their Dining-Hall-Swipe Data," *Chronicle of Higher Education*, April 9, 2019, https://www.chronicle.com/article/are-students-socially-connected-check-their-dining-hall-swipe-data/.

29. Dawn Moore, "The Benevolent Watch: Therapeutic Surveillance in Drug Court," *Theoretical Criminology* 15, no. 3 (2011): 257, https://doi.org/10.1177/1362480610396649.

30. Gardner, "Students under Surveillance?"

31. Quoted in Gardner, "Students under Surveillance?"

32. Predi Analytics, "Services," accessed October 1, 2023, https://www.predianalytics.com/services; Krystal L. Williams and BreAnna L. Davis,

"Public and Private Investments and Divestments in Historically Black Colleges and Universities," *American Council of Education Issue Brief* (2019): 1–10, https://vtechworks.lib.vt.edu/bitstream/handle/10919/89184/PublicPrivateHbcus.pdf.

33. Tobias Fiebig, Seda Gürses, Carlos H. Gañán, Erna Kotkamp, Fernando Kuipers, Martina Lindorfer, Menghua Prisse, and Taritha Sari, "Heads in the Clouds: Measuring the Implications of Universities Migrating to Public Clouds," *arXiv* (2021): 7, https://arxiv.org/abs/2104.09462/.

34. Fiebig et al., "Heads in the Clouds."

35. Blackboard, "Enhance Your Student Success Strategy," last modified 2017, https://www.blackboard.com/resources/enhance-your-student-success-strategy/.

36. Anthology, "Blackboard Learn," accessed October 19, 2023, https://www.anthology.com/products/teaching-and-learning/learning-effectiveness/blackboard-learn/.

37. Fiebig et al., "Heads in the Clouds."

38. Janja Komljenovic, "The Rise of Education Rentiers: Digital Platforms, Digital Data and Rents," *Learning, Media and Technology* 46, no. 3 (June 2020): 230–332, https://doi.org/10.1080/17439884.2021.1891422.

39. Alan Rubel and Kyle M. L. Jones, "Student Privacy in Learning Analytics: An Information Ethics Perspective," *Information Society* 32, no. 2, 143–159, https://doi.org/10.1080/01972243.2016.1130502.

40. Marko Teräs, Juha Suoranta, Hanna Teräs, and Mark Curcher, "Post-COVID-19 Education and Education Technology 'Solutionism': A Seller's Market," *Postdigital Science and Education* 2 (July 2020): 870, https://doi.org/10.1007/s42438-020-00164-x.

41. Matthew Roza, "Student Fear for Their Data Privacy after University of California Invests in Private Equity Firm," *Salon*, July 28, 2020, https://www.salon.com/2020/07/28/students-fear-for-their-data-privacy-after-university-of-california-invests-in-private-equity-firm/.

42. Roxana Marachi and Lawrence Quill, "The Case of Canvas: Longitudinal Datafication through Learning Management Systems," *Teaching in Higher Education* 25, no. 4 (2020): 424, https://doi.org/10.1080/13562517.2020.1739641.

43. Kwet and Prinsloo, "The 'Smart' Classroom," 519.

44. Kimberly Williamson and Rene F. Kizilcec, "A Review of Learning Analytics Dashboard Research in Higher Education: Implications for Justice, Equity, Diversity, and Inclusion," *LAK22: 12th International Learning Analytics and Knowledge Conference (Association for Computing Machinery)* (2022): 260–270, https://doi.org/10.1145/3506860.3506900.

45. Elana Zeide, "The Limits of Education Purpose Limitations," *University of Miami Law Review* 71, no. 2 (2017): 524–525, https://repository.law.miami.edu/umlr/vol71/iss2/8.

46. EduNav, "The EduNav Suite," accessed October 1, 2023, https://edunav.com/.

47. EduNav, "EduNav SmartPlan," accessed October 1, 2023, https://edunav.com/edunav-smartplan/.

48. Madisson Whitman, "'We Called That a Behavior': The Making of Institutional Data," *Big Data & Society* 7, no. 1 (2020): 6, https://doi.org/10.1177/2053951720932200.

49. Sasha Costanza-Chock, *Design Justice: Community-Led Practices to Build the Worlds We Need* (Cambridge, MA: MIT Press, 2020).

50. Alex Hanna, Emily Denton, Andrew Smart, and Jamilia Smith-Loud, "Towards a Critical Race Methodology in Algorithmic Fairness," in *Proceedings of the 2020 Conference on Fairness, Accountability, and Transparency* (2020): 501–512, https://doi.org/10.48550/arXiv.1912.03593.

51. W. Carson Byrd, *Behind the Diversity Numbers: Achieving Racial Equity on Campus* (Boston: Harvard Education Press, 2021). 132.

52. Doty, "Library Analytics as Moral Dilemmas," 3.

53. James P. Walsh, "Education or Enforcement? Enrolling Universities in the Surveillance and Policing of Migration" *Crime, Law, and Social Change* 71 (2019): 335, https://doi.org/10.1007/s10611-018-9792-9.

54. Chris Gilliard, "How Ed Tech Is Exploiting Students," *Chronicle of Higher Education*, April 8, 2018, https://www.chronicle.com/article/how-ed-tech-is-exploiting-students/.

55. Morten Hansen and Janja Komlijenovic, "Automating Learning Situations in EdTech: Techno-Commercial Logic of Assetisation," *Postdigital Science and Education* 5, no. 1 (2022): 109, https://doi.org/10.1007/s42438-022-00359-4.

56. Kwet and Prinsloo, "The 'Smart' Classroom," 520.

57. Mitch Daniels, "Someone Is Watching You," *Washington Post*, March 27, 2018, https://www.washingtonpost.com/opinions/its-okay-to-be-paranoid-someone-is-watching-you/2018/03/27/1a161d4c-2327-11e8-86f6-54bfff693d2b_story.html/.

Chapter 4. Wellness

1. Sarah Ketchen Lipson, Emily G. Lattie, and Daniel Eisenberg, "Increased Rates of Mental Health Service Utilization by U.S. College Students: 10-Year Population-Level Trends (2007–2017)," *Psychiatric Services* 70, no. 1 (2018): 60–63, https://doi.org/10.1176/appi.ps.201800332.

2. Teresa M. Evans, Lindsay Bira, Jazmin Beltran Gastelum, L Todd Weiss, and Nathan L Vanderford, "Evidence for a Mental Health Crisis in Graduate Education," *Nature Biotechnology* 36 (2018): 282–284.

3. See, for example, Sarah Ketchen Lipson, Adam Kern, Daniel Eisenberg, and Alfiee M. Breland-Noble, "Mental Health Disparities among College

Students of Color," *Journal of Adolescent Health* 63, no. 3 (2018): 348–56, https://doi.org/10.1016/j.jadohealth.2018.04.014; Sarah Ketchen Lipson, Julia Raifman, Sara Abelson, and Sari L. Reisner, "Gender Minority Mental Health in the U.S.: Results of a National Survey on College Campuses," *American Journal of Preventative Medicine* 57, no. 3 (2019): 293–301, https://doi.org/10.1016/j.amepre.2019.04.025; and Cindy H. Liu, Courtney Stevens, Sylvia H. M. Wong, Miwa Yasui, and Justin A. Chen, "The Prevalence and Predictors of Mental Health Diagnoses and Suicide among U.S. College Students: Implications for Addressing Disparities in Service Use," *Depression & Anxiety* 36, no. 1 (2019): 8–17, https://doi.org/10.1002/da.22830.

4. Sasha Costanza-Chock, "Design Justice: Towards an Intersectional Feminist Framework for Design Theory and Practice," *Proceedings of the Design Research Society* (2018): 4, https://ssrn.com/abstract=3189696.

5. David P. Kraft, "One Hundred Years of College Mental Health," *Journal of American College Health* 59, no. 6 (2011): 477.

6. Kraft, "One Hundred Years of College Mental Health."

7. Frankwood E. Williams, "Mental Hygiene and the College Student," *Mental Hygiene* 5 (1921): 301.

8. Williams, "Mental Hygiene," 300.

9. Raymond Wolters, *The New Negro on Campus: Black College Rebellions of the 1920s* (Princeton, NJ: Princeton University Press, 1975).

10. Wolters, *The New Negro on Campus*, 301.

11. Wolters, *The New Negro on Campus*, 285.

12. Sol Cohen, "The Mental Hygiene Movement, the Development of Personality and the School: The Medicalization of American Education," *History of Education Quarterly* 23, no. 2 (1983): 141.

13. David Beer, *Metric Power* (London: Palgrave Macmillan, 2016), 71.

14. Ganaele Langlois and Greg Elmer, "Impersonal Subjectivation from Platforms to Infrastructures," *Media, Culture & Society* 41, no. 2 (2018): 237, https://doi.org/10.1177/0163443718818374.

15. Michel Foucault, *The Care of the Self: The History of Sexuality*, vol. 3 (New York: Pantheon, 1986).

16. Cohen, "The Mental Hygiene Movement," 50.

17. Emily Mertz, "University of Alberta Students Get in Touch with Well-Being, Mental Health through WellTrack App," *Global News*, November 21, 2018, https://globalnews.ca/news/4686918/mental-health-university-of-alberta-students-welltrack-app/.

18. Olivia Ginsberg, "UM Counseling Center Offers New App to Promote Mental Health," *Miami Hurricane*, last modified November 5, 2018, https://www.themiamihurricane.com/2018/11/05/um-counseling-center-offers-new-app-to-promote-mental-health/.

19. Jodie Vanderslot, "Mental Health through WellTrack," *Excalibur*, last modified December 1, 2017, https://www.excal.on.ca/health/2017/12/01/mental-health-through-welltrack/.

20. Ophelie Zalcmanis-Lai, "Exclusive: University Set to Introduce Online Mental Health Platform to Students," *Ryersonian*, April 9, 2016, https://ryersonian.ca/exclusive-university-set-to-introduce-online-mental-health-platform-to-students/.

21. Michael S. Dunbar, Lisa Sontag-Padilla, Rajeev Ramchand, Rachana Seelam, and Bradley D. Stein, "Mental Health Service Utilization among Lesbian, Gay, Bisexual, and Questioning or Queer College Students," *Journal of Adolescent Health* 61, no. 3 (2017): 294–301, https://doi.org/10.1016/j.jadohealth.2017.03.008; Lipson et al., "Mental Health Disparities among College Students of Color."

22. David Ferraro, "Psychology in the Age of Austerity," *Psychotherapy and Politics International* 14, no. 1 (2016): 17–24, https://doi.org/10.1002/ppi.1369.

23. Mirai So, Sosei Yamaguchi, Sora Hashimoto, Mitsuhiro Sado, Furukawa A. Toshi, and Paul McCrone, "Is Computerised CBT Really Helpful for Adult Depression? A Meta-analytic Re-evaluation of CCBT for Adult Depression in Terms of Clinical Implementation and Methodological Validity," *BMC Psychiatry*, no. 13 (2013): 113–27, https://doi.org/10.1186/1471-244X-13-113.

24. Rachel I. Rosner, "Manualizing Psychotherapy: Aaron T. Beck and the Origins of Cognitive Therapy of Depression," *European Journal of Psychotherapy & Counselling* 20, no. 1 (2018): 43, https://doi.org/10.1080/13642537.2017.1421984.

25. Åsa Jansson, "From Self-Help to CBT: Regulating Emotion in a (Neo) liberal World," *History of Emotions Blog*, December 11, 2017, https://emotionsblog.history.qmul.ac.uk/2017/12/from-self-help-to-cbt-regulating-emotion-in-a-neoliberal-world/.

26. David Ferraro, "Psychology in the Age of Austerity," *Psychotherapy and Politics International* 14, no. 1 (2016): 9, https://doi.org/10.1002/ppi.1369.

27. Deborah Lupton, *The Quantified Self* (Cambridge: Polity Press), 50.

28. Sena Cerci, "Embodying Self-Tracking: A Feminist Exploration of Collective Meaning-Making of Self-Tracking Data" (MA thesis, Malmo University, 2018).

29. Constantine Gidaris, "Surveillance Capitalism, Datafication, and Unwaged Labour: The Rise of Wearable Fitness Devices and Interactive Life Insurance," *Surveillance & Society* 17, nos. 1/2 (2019): 132–138, https://doi.org/10.24908/ss.v17i1/2.12913.

30. Phoebe V. Moore, *The Quantified Self in Precarity: Work, Technology and What Counts* (New York: Routledge, 2018), 3.

31. Simone Browne, *Dark Matters: On the Surveillance of Blackness* (Durham, NC: Duke University Press, 2015); Dorothy E. Roberts, "Book Review: Digitizing the Carceral State," *Harvard Law Review* 132, no. 6 (2019): 1695–1729, https://www.jstor.org/stable/26800052; Ruha Benjamin, *Race after Technology: Abolitionist Tools for a New Jim Code* (Cambridge: Polity, 2019).

32. Shaun L. Gabbidon, "Racial Profiling by Store Clerks and Security Personnel in Retail Establishments: An Exploration of 'Shopping While Black,'" *Journal of Contemporary Criminal Justice* 19, no. 3 (2003): 345–64, https://doi.org/10.1177/1043986203254531.

33. Adrienne Green, "The Cost of Balancing Academia and Racism," *Atlantic*, January 21, 2016, https://www.theatlantic.com/education/archive/2016/01/balancing-academia-racism/424887/.

34. Ebony O. McGee and David Stovall, "Reimagining Critical Race Theory in Education: Mental Health, Healing, and the Pathway to Liberatory Praxis," *Educational Theory* 65, no. 5 (2015): 491–511, https://doi.org/10.1111/edth.12129.

35. McGee and Stovall, "Reimagining Critical Race Theory," 501.

36. Alicia Hahn, "2023 Student Loan Debt Statistics: Average Student Loan Debt," *Forbes*, last modified July 16, 2023, https://www.forbes.com/advisor/student-loans/average-student-loan-debt-statistics/.

37. Anna Helhoski, "Why the Student Debt Crisis Hits Black Borrowers Harder," *Nerd Wallet*, last modified February 7, 2018, https://www.nerdwallet.com/blog/loans/student-loans/black-student-debt-crisis/.

38. Helhoski, "Why the Student Debt Crisis Hits Black Borrowers Harder."

39. See, for example, Debt Collective, "End Student Debt!," last accessed January 25, 2021, https://strike.debtcollective.org/.

40. Audre Lorde, *A Burst of Light and Other Essays* (London: Sheba Feminist, 1988), 130.

41. Sara Ahmed, "Selfcare as Warfare," *Feministkilljoys* (blog), August 25, 2014, https://feministkilljoys.com/2014/08/25/selfcare-as-warfare/.

42. Moore, *The Quantified Self in Precarity*, 119.

43. Deborah Lupton, "The Diverse Domains of Quantified Selves: Self-Tracking Modes and Dataveillance," *Economy and Society* 45, no. 1 (2016): 103, https://doi.org/10.1080/03085147.2016.1143726.

44. CyberPsyc Software Solutions, "Privacy Policy," accessed January 31, 2019, https://app.welltrack.com/privacy-policy.

45. Frank Pasquale, *The Black Box Society: The Secret Algorithms That Control Money and Information* (Cambridge, MA: Harvard University Press, 2015).

46. Alexandra S. Levine, "Suicide Hotline Shares Data with For-Profit Spinoff, Raising Ethical Questions," *Politico*, January 28, 2022, https://www

.politico.com/news/2022/01/28/suicide-hotline-silicon-valley-privacy-debates-00002617.

47. Chris Gilliard, "How Ed Tech Is Exploiting Students," *Chronicle of Higher Education*, April 8, 2018, https://www.chronicle.com/article/how-ed-tech-is-exploiting-students/.

48. CyberPsyc Software Solutions, "Privacy Policy," last accessed January 31, 2019, https://app.welltrack.com/privacy-policy.

49. Les Hutchinson and Maria Novotny, "Teaching a Critical Literacy of Wearables: A Feminist Surveillance as Care Pedagogy," *Computers and Composition* 50 (December 2018): 106, https://doi.org/10.1016/j.compcom.2018.07.006.

50. CyberPsyc Software Solutions, "Privacy Policy."

51. Rosalind Gill and Ngaire Donaghue, "Resilience, Apps and Reluctant Individualism: Technologies of Self in the Neoliberal Academy," *Women's Studies International Forum* 54 (January/February 2016): 98, https://doi.org/10.1016/j.wsif.2015.06.016.

52. WellTrack, "Clinicians & Administrators," accessed January 30, 2019, https://welltrack.com/for-higher-ed/.

53. Quoted in Ali Montag, "This University Is Putting Amazon Echo Speakers in Every Dorm Room," CNBC, August 21, 2018, https://www.cnbc.com/2018/08/21/this-university-is-putting-amazon-echo-speakers-in-every-dorm-room.html.

54. Dalvin Brown, "Alexa Goes to College: Echo Dots Move into Dorms on Campus," *USA Today*, September 6, 2018, https://www.usatoday.com/story/money/2018/09/06/college-students-echo-dots-dorm-rooms/1087251002/.

55. "'Alexa, Why Are You Here?,'" *Mercury*, April 29, 2019, https://utdmercury.com/alexa-where-are-you-here/.

56. Saint Louis University, "Welcome to SLU, Alexa," YouTube video, posted by SaintLouisUniversity, August 9, 2018, https://youtu.be/Q8jfcxMYbQw.

57. Somen Saha, "World's First Voice-Enabled Student Experience by N-Powered.com," YouTube video, January 30, 2018, https://youtu.be/wtdroQzWRhY.

58. Laura Hirsch and Michelle Castillo, "Amazon Has Big Plans for Alexa Ads in 2018: It's Discussing Options with P&G, Clorox, and Others," CNBC, January 2, 2018, https://www.cnbc.com/2018/01/02/amazon-alexa-is-opening-up-to-more-sponsored-product-ads.html.

59. Ethan Sacks, "Alexa Privacy Fail Highlights Risks of Smart Speakers," NBC News, May 26, 2018, https://www.nbcnews.com/tech/innovation/alexa-privacy-fail-highlights-risks-smart-speakers-n877671.

60. Sean Hollister, "Today I Learned Amazon Has a Form So Police Can Get My Data without Permission or a Warrant," *Verge*, July 15, 2022, https://www.theverge.com/2022/7/14/23219419/amazon-ring-law-enforcement-no-warrant-no-consent.

61. Montag, "This University."

62. Jonathan Crary, *24/7: Late Capitalism and the Ends of Sleep* (London: Verso, 2014), 46.

63. WellTrack, "Clinicians & Administrators."

64. Caroline Simon, "More and More Students Need Mental Health Services: But Colleges Struggle to Keep Up," *USA Today*, May 4, 2017, https://www.usatoday.com/story/college/2017/05/04/more-and-more-students-need-mental-health-services-but-colleges-struggle-to-keep-up/37431099/.

65. Steve Jenkins (sales development representative for WellTrack), email to author, June 20, 2019.

66. Jenkins, email to author.

67. Kristiina Brunila and Leena-Maija Rossi, "Identity Politics, the Ethos of Vulnerability, and Education," *Educational Philosophy and Theory* 50, no. 3 (2017): 287–298, https://doi.org/10.1080/00131857.2017.1343115.

68. Simon, "More and More Students."

69. Sasha Costanza-Chock, *Design Justice: Community-Led Practices to Build the Worlds We Need* (Cambridge, MA: MIT Press, 2020).

70. Benjamin, *Race after Technology*, 176.

71. Jeremy Knox, Ben Williamson, and Sian Bayne, "Machine Behaviourism: Future Vision of 'Learnification' and 'Datafication' across Human and Digital Technologies," *Learning, Media and Technology* 45, no. 1 (2020): 42, https://doi.org/10.1080/17439884.2019.1623251.

Chapter 5. Security

1. Keith Fraser, "Software Company Sues UBC Employee over Tweets Involving Confidential Videos," *Province*, September 3, 2020, https://theprovince.com/news/software-company-sues-ubc-employee-over-tweets-involving-confidential-videos/wcm/551266ea-a842-4a48-a8e1-1f3fb732f911/.

2. Shea Swauger, "Software That Monitors Students during Tests Perpetuates Inequality and Violates Their Privacy," *MIT Technology Review*, August 7, 2020, https://www.technologyreview.com/2020/08/07/1006132/software-algorithms-proctoring-online-tests-ai-ethics/.

3. Georgia Yee, "Protect Student Privacy: A Renewed Call to Action against Proctorio," last modified October 1, 2020, https://docs.google.com/document/d/1117835S2RQkQN_-Ij8nZ2qEzCKgNWqle-O7onKdzpaA/edit.

4. Shea Swauger, "Our Bodies Encoded: Algorithmic Test Proctoring in Higher Education," *Hybrid Pedagogy*, April 2, 2020, https://hybridpedagogy.org/our-bodies-encoded-algorithmic-test-proctoring-in-higher-education/.

5. Swauger, "Software That Monitors Students."

6. Monica Chin, "College Student Sues Proctorio after Source Code Copyright Claim," *Verge*, May 5, 2021, https://www.theverge.com/2021/4/22/22397499/proctorio-lawsuit-electronic-frontier-foundation-test-proctoring-software.

7. Thierry Balzacq, preface to *Securitization Theory: How Security Problems Emerge and Dissolve*, ed. Thierry Balzacq (London: Routledge, 2011), xiii.

8. Proctorio, "Online Proctoring," accessed May 19, 2021, https://proctorio.com/products/online-proctoring.

9. Ann Gibson Winfield, *Eugenics and Education in America: Institutionalized Racism and the Implications of History, Ideology, and Memory* (New York: Peter Lang, 2007).

10. Jack Schneider and Ethan Hutt, "Making the Grade: A History of the A–F Marking Scheme," *Journal of Curriculum Studies* 46, no. 2 (2014): 218, http://dx.doi.org/10.1080/00220272.2013.790480.

11. Marianne Madoré, Anna Zeemont, Joaly Burgos, Jane Guskin, Hailey Lam, and Andréa Stella, "Resisting Surveillance, Practicing/Imagining the End of Grading," *Journal of Interactive Technology & Pedagogy*, December 10, 2021, https://jitp.commons.gc.cuny.edu/resisting-surveillance-practicing-imagining-the-end-of-grading/.

12. Lincoln Taiz, Joel Yellin, Deanna Shemek, Karen Bassi, and Tony Tromba, "UCSC Admission: History and Opportunities," accessed May 19, 2021, https://senate.ucsc.edu/archives/Past%20Issues/narrative-evaluations/AdmissionsHisandOpp_doc.pdf.

13. Emma Bowman, "Here's What Happened when Affirmative Action Ended at California Public Colleges," NPR, June 30, 2023, https://www.npr.org/2023/06/30/1185226895/heres-what-happened-when-affirmative-action-ended-at-california-public-colleges.

14. Monica Crespo, "Fair, Transparent, and Accountable Algorithmic Decision-Making: What Is the Role of Human-in-the-Loop?," *iSChannel* 17, no. 1 (2022): 28–41, https://ischannel.lse.ac.uk/articles/208/.

15. Madoré et al., "Resisting Surveillance, Practicing/Imagining the End of Grading."

16. Sarah Myers West, Meredith Whittaker, and Kate Crawford, "Discriminating Systems: Gender, Race, and Power in AI," AI Now (2019), https://ainowinstitute.org/publication/discriminating-systems-gender-race-and-power-in-ai-2.

17. Louise Seamster and Raphaël Charron Chénier, "Predatory Inclusion and Education Debt: Rethinking the Racial Wealth Gap," *Social Currents* 4, no. 3 (2017): 199–207, https://doi.org/10.1177/2329496516686620.

18. Swauger, "Our Bodies Encoded."

19. Charles Logan, "Towards Abolishing Online Proctoring: Counter-Narratives, Deep Change, and Pedagogies of Educational Dignity," *Journal of Interactive Technology & Pedagogy*, December 2021, https://jitp.commons.gc.cuny.edu/toward-abolishing-online-proctoring-counter-narratives-deep-change-and-pedagogies-of-educational-dignity/.

20. Xiaolin Wu and Xi Zhang, "Automated Inference on Criminality Using Face Images," *arXiv* (2016): 1–9, https://arxiv.org/pdf/1611.04135v1.pdf.

21. Raja Saravanan, "Facial Recognition Can Give Students Better Service (and Security)," *Ellucian* (blog), accessed May 19, 2023, https://www.ellucian.com/blog/facial-recognition-campus-benefits-security-risks.

22. Fight for the Future and Students for Sensible Drug Policy, "Stop Facial Recognition on Campus," accessed May 19, 2023, https://www.banfacialrecognition.com/campus/.

23. Saravanan, "Facial Recognition Can Give Students Better Service."

24. Jay W. Goff and Christopher M. Shaffer, "Big Data's Impact on College Admission Practices and Recruitment Strategies," in *Building a Smarter University: Big Data, Innovation, and Analytics*, ed. Jason Lane (Albany: State University of New York Press, 2014), 93–120.

25. "Technology Fosters Deeper Campus Connections," *Ellucian* (blog), accessed May 19, 2023, https://www.ellucian.com/blog/technology-fosters-deeper-campus-connections.

26. "Technology Fosters Deeper Campus Connections."

27. Coherent Market Insights, "Emotion Detection and Recognition Market to Surpass US$ 40.5 Billion by 2030," *Global Newswire*, March 27, 2023, https://www.globenewswire.com/news-release/2023/03/27/2634806/0/en/Emotion-Detection-and-Recognition-Market-to-surpass-US-40-5-Billion-by-2030-Says-Coherent-Market-Insights-CMI.html.

28. Lauren Rhue, "Racial Influence on Automated Perceptions of Emotions," *Social Science Research Network*, November 2018, 1–11, https://papers.ssrn.com/sol3/papers.cfm?abstract_id=3281765.

29. Amy G. Halberstadt, Vanessa L. Castro, Qiao Chu, Fantasy T. Lozada, and Calvin M. Sims, "Preservice Teachers' Racialized Emotion Recognition, Anger Bias, and Hostility Attributions," *Contemporary Educational Psychology* 54 (July 2018): 125–138, https://doi.org/10.1016/j.cedpsych.2018.06.004.

30. Kari Paul, "'Ban This Technology': Students Protest US Universities' Use of Facial Recognition," *Guardian*, March 2, 2020, https://www.theguardian.com/us-news/2020/mar/02/facial-recognition-us-colleges-ucla-ban.

31. Claire Galligan, Hannah Rosenfeld, Molly Kleinman, and Shobita Parthasarathy, "Cameras in the Classroom: Facial Recognition Technology in Schools," *Technology Assessment Project Report* (2020): 1–115, https://stpp.fordschool.umich.edu/sites/stpp/files/uploads/file-assets/cameras_in_the_classroom_full_report.pdf.

32. Abigail Boggs, "On Borders and Academic Freedom: Noncitizen Students and the Limits of Rights," *AAUP Journal of Academic Freedom* 11 (2020): 6.

33. Boggs, "On Borders and Academic Freedom," 11.

34. Harrison Young, "UGA Dining Halls to Introduce Iris Scanners," *Red & Black*, last modified April 22, 2017, https://www.redandblack.com/uganews/uga-dining-halls-to-introduce-eye-scanners/article_0b1703e2-2562-11e7-9dac-f38fb6755209.html.

35. Katherine Lippert, "Amid Coronavirus, USC Is Requiring Facial Recognition Scans of Students Living on Campus, but the Technology Sparks Controversy," *USC Annenberg Media*, last modified May 15, 2020, https://www.uscannenbergmedia.com/2020/05/15/amid-coronavirus-usc-is-requiring-facial-recognition-scans-of-students-living-on-campus-but-the-technology-sparks-controversy/.

36. Lindsay McKenzie, "Secure File Sharing Compromises University Security," *Inside Higher Ed*, last modified April 6, 2021, https://www.insidehighered.com/news/2021/04/07/accellion-data-security-breach-latest-hit-universities.

37. EDUCAUSE, "Higher Education Community Vendor Assessment Toolkit," last modified December 17, 2021, https://library.educause.edu/resources/2020/4/higher-education-community-vendor-assessment-toolkit.

38. EDUCAUSE, accessed May 20, 2023, https://www.educause.edu/.

39. EDUCAUSE, "CAUSE History," accessed October 17, 2023, https://www.educause.edu/about/mission-and-organization/our-history/cause-history.

40. Arun Kundnani and Deepa Kumar, "Race, Surveillance, and Empire," *International Socialist Review*, no. 96 (2016), https://isreview.org/issue/96/race-surveillance-and-empire.

41. Jennifer Doyle, *Campus Sex, Campus Security* (Cambridge: Semiotext(e), 2015), 16.

42. Davarian L. Baldwin, *In the Shadow of the Ivory Tower: How Universities are Plundering Our Cities* (New York: Bold Type Books, 2021), 182.

43. John J. Sloan III and Bonnie S. Fisher, *The Dark Side of the Ivory Tower: Campus Crime as a Social Problem* (Cambridge: Cambridge University Press, 2010).

44. Rebecca Dolinsky Graham and Amanda Konradi, "Contextualizing the 1990 Campus Security Act and Campus Sexual Assault in Intersectional and Historical Terms," *Journal of Aggression, Conflict and Peace Research* 10, no. 2 (2018): 96.

45. Graham and Konradi, "Contextualizing the 1990 Campus Security Act," 98.

46. Sloan and Fisher, *The Dark Side of the Ivory Tower*, 115.

47. Kelly Nielsen, Laura T. Hamilton, and Veronica Lerma, "Policing College Campuses: The Production of Racialized Risk," *UF Law Faculty Publications* (2020): 6, https://doi.org/10.31235/osf.io/pe8my.

48. Lippert, "Amid Coronavirus, USC Is Requiring Facial Recognition Scans."

49. John Jay Sloan III, "Race, Violence, Justice, and Campus Police," *Footnotes: A Publication of the American Sociological Association* 48, no. 4 (2020): 9, https://www.asanet.org/news-events/footnotes/jul-aug-2020/features/race-violence-justice-and-campus-police.

50. Nielsen, Hamilton, and Lerma, "Policing College Campuses," 3–7.

51. Jay Stanley, "Four Problems with the ShotSpotter Gunshot Detection System," ACLU, last modified October 14, 2021, https://www.aclu.org/news/privacy-technology/four-problems-with-the-shotspotter-gunshot-detection-system; Georgia Gee, "Un-alarmed: AI Tries (and Fails) to Detect Weapons in Schools," *Intercept*, May 7, 2023, https://theintercept.com/2023/05/07/ai-gun-weapons-detection-schools-evolv/.

52. Nielsen, Hamilton, and Lerma, "Policing College Campuses," 4.

53. Dave Maass, "Scholars under Surveillance: How Campus Police Use High Tech to Spy on Students," Electronic Frontier Foundation, last modified March 9, 2021, https://www.eff.org/deeplinks/2021/03/scholars-under-surveillance-how-campus-police-use-high-tech-spy-students#social.

54. Baldwin, *In the Shadow of the Ivory Tower*.

55. Mass, "Scholars under Surveillance."

56. Social Sentinel, "Our Solution," accessed May 18, 2020, https://navigate360.com/solutions/social-sentinel-scanning-businesses/.

57. Marisa R. Randazzo and J. Kevin Cameron, "From Presidential Protection to Campus Security: A Brief History of Threat Assessment in North American Schools and Colleges," *Journal of College Student Psychotherapy* 26, no. 4 (2012): 277–290, https://doi.org/10.1080/87568225.2012.711146.

58. John J. Sloan III, "The Modern Campus Police: An Analysis of their Evolution, Structure, and Function," *American Journal of Police* 11, no. 1 (1992): 86; Grace Watkins, "'Cops Are Cops': American Campus Police and the Global Carceral Apparatus," *Comparative American Studies* 19, nos. 3/4 (2020): 243, https://doi.org/10.1080/14775700.2021.1895039.

59. Philip Lee, "The Curious Life of in Loco Parentis in American Universities," *Higher Education in Review* 8 (2011): 67, https://scholar.harvard.edu/files/philip_lee/files/vol8lee.pdf.

60. Corbretti D. Williams, "Race and Policing in Higher Education," *Activist History Review*, November 19, 2019, https://activisthistory.com/2019/11/19/race-and-policing-in-higher-education/.

61. Michael Clay Smith, "The Ancestry of Campus Violence," *New Directions for Student Services* 1989, no. 4 (1989): 8, https://doi.org/10.1002/ss.37119894703.

62. Sloan, "The Modern Campus Police," 88.

63. White v. Davis, vol. 13, Cal. 3d, March 24, 1975, https://law.justia.com/cases/california/supreme-court/3d/13/757.html.

64. Kundnani and Kumar, "Race, Surveillance, and Empire."

65. Watkins, "'Cops Are Cops,'" 251.

66. Watkins, "'Cops Are Cops,'" 251.

67. James Wilkinson, "UC Davis Paid at Least $175,000 to Hide References to Infamous 2011 'Pepper-Spray Incident' from Google Searches," *Daily Mail*, last modified April 14, 2016, https://www.dailymail.co.uk/news/article-3540299/UC-Davis-chancellor-Lind-Katehi-fire-taxpayer-funded-uni-paid-175-000-hide-references-infamous-2011-pepper-spray-incident-Google-searches-university.html.

68. Farah Godrej, "Neoliberalism, Militarization, and the Price of Dissent: Policing Protest at the University of California," in *The Imperial University: Academic Repression and Scholarly Dissent*, ed. Piya Chatterjee and Sunaina Maira (Minneapolis: University of Minnesota Press, 2014), 135.

69. Dave Maass and Mike Katz-Lacabe, "Alameda and Contra Costa County Sheriffs Flew Drones Over Protests," Electronic Frontier Foundation, last modified December 5, 2018, https://www.eff.org/deeplinks/2018/12/alameda-and-contra-costa-county-sheriffs-flew-drones-over-protests; Sidney Fusell, "Did a University Use Facial Recognition to ID Student Protestors?," *Wired*, last modified November 18, 2020, https://www.wired.com/story/did-university-use-facial-recognition-id-student-protesters/.

70. Gabe Evans, Nick Mitchell, and Taylor Wondergem, "Scenes from the Wildcat Strike: A Documentary History," *Critical Ethnic Studies* 6, no. 2 (2021), https://doi.org/10.5749/CES.0602.wildcat-strike.

71. Nicole Karlis, "Emails Show UC Santa Cruz Police Used Military Surveillance to Suppress Student Strike," *Salon*, last modified May 18, 2020, https://www.salon.com/2020/05/18/emails-show-uc-santa-cruz-police-used-military-surveillance-to-suppress-grad-student-strike/.

72. Evans, Mitchell, and Wondergem, "Scenes from the Wildcat Strike."

73. Evans, Mitchell, and Wondergem, "Scenes from the Wildcat Strike."

74. Stephanie Weagle, "From Safe Campus to Smart Campus," *Campus Security Today*, last modified April 1, 2019, https://campuslifesecurity.com/Articles/2019/04/01/From-Safe-Campus-to-Smart-Campus.aspx?Page=1.

75. Watkins, "'Cops Are Cops,'" 243.

76. Watkins, "'Cops Are Cops,'" 243.

77. Stuart Shrader, *Badges without Borders: How Global Counterinsurgency Transformed American Policing* (Berkeley: University of California Press, 2019).

78. Watkins, "'Cops Are Cops,'" 246.

79. Watkins, "'Cops Are Cops,'" 252.

80. Godrej, "Neoliberalism, Militarization, and the Price of Dissent," 125.

81. Dylan Rodríguez, "Beyond 'Police Brutality': Racist State Violence and the University of California," *American Quarterly* 64, no. 2 (2012): 311, https://doi.org/10.1353/aq.2012.0012.

82. Abigail Boggs and Nick Mitchell, "Critical University Studies and the Crisis Consensus," *Feminist Studies* 44, no. 2 (2018): 434–435, https://doi.org/10.15767/feministstudies.44.2.0432.

83. Donna Haraway, "Situated Knowledges: The Science Question in Feminism and the Privilege of Partial Perspective," *Feminist Studies* 14, no. 3 (1988): 575–599, https://doi.org/10.2307/3178066.

84. Boggs and Mitchell, "Critical University Studies and the Crisis Consensus," 451.

85. Craig Steven Wilder, *Ebony & Ivy: Race, Slavery, and the Troubled History of America's Universities* (New York: Bloomsbury Press, 2013).

86. Wilder, *Ebony & Ivy*; Robert Lee and Tristan Ahtone, "Land-Grab Universities," *High Country News*, March 30, 2020, https://www.hcn.org/issues/52.4/indigenous-affairs-education-land-grab-universities; Manan Ahmed Asif, "Technologies of Power—from Area Studies to Data Sciences," *Spheres: Journal for Digital Cultures*, last modified November 20, 2019, http://spheres-journal.org/technologies-of-power-from-area-studies-to-data-sciences/.

87. Richard M. Abrams, "The US Military and Higher Education: A Brief History," *Annals of the American Academy* 502, no. 1 (1989): 15–28, https://doi.org/10.1177/0002716289502001002.

88. Piya Chatterjee and Sunaina Maira, "The Imperial University: Race, War, and the Nation-State," in *The Imperial University: Academic Repression and Scholarly Dissent*, ed. Piya Chatterjee and Sunaina Maira (Minneapolis: University of Minnesota Press, 2014), 13.

89. Asif, "Technologies of Power."

90. Abrams, "The US Military and Higher Education," 28.

91. Asif, "Technologies of Power."

92. Indigo Oliver, "Inside Lockheed Martin's Sweeping Recruitment on College Campuses," *In These Times*, August 11, 2022, https://inthesetimes.com/article/lockheed-martin-recruiting-military-industrial-complex-student-debt; Rodrigo Ochigame, "The Invention of 'Ethical AI': How Big Tech Manipulates Academia to Avoid Regulation," *Intercept*, December 20, 2019, https://theintercept.com/2019/12/20/mit-ethical-ai-artificial-intelligence/.

93. Amy Zegart and Michael Morell, "Spies, Lies, and Algorithms: Why US Intelligence Agencies Must Adapt or Fail," *Foreign Affairs*, April 16, 2019, https://www.foreignaffairs.com/articles/2019-04-16/spies-lies-and-algorithms.

94. Lindsay Weinberg, "Rethinking Fairness: An Interdisciplinary Survey of Critiques of Hegemonic Fairness Approaches," *Journal of Artificial Intelligence Research* 74 (2022): 78–79, https://doi.org/10.1613/jair.1.13196.

95. Meredith Whittaker, "The Steep Cost of Capture," *Interactions* 18, no. 6 (2021): 50–55, https://doi.org/10.1145/3488666.

96. Whittaker, "The Steep Cost of Capture," 54.

97. Sidney Fussell, "You No Longer Own Your Face," *Atlantic*, June 27, 2019, https://www.theatlantic.com/technology/archive/2019/06/universities-record-students-campuses-research/592537/.

98. Adam Harvey and Jules LaPlace, "UnConstrained College Students Dataset," accessed February 4, 2024, https://exposing.ai/uccs/.

99. Joy Bualamwini and Timnit Gebru, "Gender Shades: Intersectional Accuracy Disparities in Commercial Gender Classification," in *Proceedings of the First Conference on Fairness, Accountability and Transparency (Proceedings of Machine Learning Research)* 81 (2018): 77–91, http://proceedings.mlr.press/v81/buolamwini18a.html; and Os Keyes, "The Misgendering Machines: Trans/HCI Implications of Automatic Gender Recognition," *Proceedings of the ACM on Human-Computer Interaction* 2 (2018), https://doi.org/10.1145/3274357.

100. Simone Browne, *Dark Matters: On the Surveillance of Blackness* (Durham, NC: Duke University Press, 2015).

101. Rachel Dubrofsky and Shoshana Amielle Magnet, introduction to *Feminist Surveillance Studies*, ed. Rachel Dubrofsky and Shoshana Amielle Magnet (Durham, NC: Duke University Press, 2015), 1–18.

102. Zach Blas and Jacob Gaboury, "Biometrics and Opacity: A Conversation," *Camera Obscura* 31, no. 2 (2016): 155–165.

103. Jasmine R. Lindabary and Danielle J. Corple, "Privacy for Whom? A Feminist Intervention in Online Research Practice," *Information, Communication & Society* 22, no. 10 (2019): 1458–1459, https://doi.org/10.1080/1369118X.2018.1438492.

104. Os Keyes, Nikki Stevens, and Jacqueline Wernimont, "The Government Is Using the Most Vulnerable People to Test Facial Recognition Software," *Slate*, March 17, 2019, https://slate.com/technology/2019/03/facial-recognition-nist-verification-testing-data-sets-children-immigrants-consent.html.

105. Cheryl Cooky, Jasmine R. Lindabary, and Danielle J. Corple, "Navigating Big Data Dilemmas: Feminist Holistic Reflexivity in Social Media Research," *Big Data & Society* 5, no. 2 (2018): 1–12, https://doi.org/10.1177/2053951718807731.

106. John Gilliom, *Overseers of the Poor: Surveillance, Resistance, and the Limits of Privacy* (Chicago: University of Chicago Press, 2001).

107. J. Adrian Stanley, "UCCS Secretly Photographed Students to Advance Facial Recognition Technology," *Colorado Springs Independent*, last modified October 15, 2020, https://www.csindy.com/temporary_news/uccs-secretly-photographed-students-to-advance-facial-recognition-technology/article_f314793e-20aa-586c-a0e9-db2cadd394a4.html.

108. Julia Angwin, Jeff Larson, Surya Mattu, and Lauren Kirchner, "Machine Bias: There's Software Used across the County to Predict Future Criminals. And It's Biased against Blacks," *ProPublica*, May 23, 2016, https://www.propublica.org/article/machine-bias-risk-assessments-in-criminal-sentencing.

109. Ben Green, "'Fair' Risk Assessments: A Precarious Approach for Criminal Justice Reform," *5th Workshop on Fairness, Accountability, and Transparency in Machine Learning* (2018): 1–5, https://scholar.harvard.edu/files/bgreen/files/18-fatml.pdf.

110. Ben Green, "'Fair' Risk Assessments," 2.

111. Megan Stevenson and Sandra G. Mayson, "The Scale of Misdemeanor Justice," *Boston University Law Review* 98, no. 731 (2018): 769–770, https://scholarship.law.upenn.edu/faculty_scholarship/2391.

112. Coalition for Critical Technology, "Abolish the #TechtoPrisonPipeline," last modified June 23, 2020, https://medium.com/@CoalitionForCriticalTechnology/abolish-the-techtoprisonpipeline-9b5b14366b16.

113. James Merricks White, "Anticipatory Logics of the Smart City's Global Imaginary," *Urban Geography* 37, no. 4 (2016): 572–589, https://doi.org/10.1080/02723638.2016.1139879.

114. Simone Tulumello and Fabio Iapaolo, "Policing the Future, Disrupting Urban Policy Today: Predictive Policing, Smart City, and Urban Policy in Memphis (TN)," *Urban Geography* 43, no. 3 (2022): 449, https://doi.org/10.1080/02723638.2021.1887634.

115. Elena Delavega and Gregory M. Blumenthal, "2023 Memphis Poverty Fact Sheet," accessed October 20, 2023, https://www.memphis.edu/socialwork/research/2023-poverty-fact-sheet-final.pdf.

116. Johanna Leigh, Sarah Dunnett, and Lisa Jackson, "Predictive Police Patrolling to Target Hotspots and Cover Response Demand," *Annals of Operations Research* 283 (2019): 395–410, https://doi.org/10.1007/s10479-017-2528-x.

117. Tulumello and Iapaolo, "Policing the Future," 452.

118. Kristian Lum and William Isaac, "To Predict and Serve?," *Significance* 13, no. 5 (2016): 14–19, https://doi.org/10.1111/j.1740-9713.2016.00960.x.

119. Cited in Lum and Isaac, "To Predict and Serve?," 18.

120. Ruha Benjamin, *Race after Technology; Abolitionist Tools for the New Jim Code* (Cambridge: Polity, 2019).

121. Tulumello and Iapaolo, "Policing the Future."

122. Coalition for Critical Technology, "Abolish the #TechtoPrisonPipeline."

123. Rashida Richardson, Jason Schultz, and Kate Crawford, "Dirty Data, Bad Predictions: How Civil Rights Violations Impact Police Data, Predictive Policing Systems, and Justice," *New York University Law Review Online* 94, no. 192 (2019): 192–233, https://papers.ssrn.com/sol3/papers.cfm?abstract_id =3333423.

124. Andrew Guthrie Ferguson, "Policing Predictive Policing," *Washington University Law Review* 94, no. 5 (2017): 1155, https://papers.ssrn.com/sol3 /papers.cfm?abstract_id=2765525.

125. Dorothy E. Roberts, "Book Review: Digitizing the Carceral State," *Harvard Law Review* 132, no. 6 (2019): 1695–1729, https://www.jstor.org/stable /26800052.

126. See Angela Davis, *Are Prisons Obsolete?* (New York: Seven Stories Press, 2023); and Shrader, *Badges without Borders*.

127. Bella Nadler, "Students Call for UCLA Administration to Abolish Campus Police Presence," *Fem Magazine*, June 6, 2021, https://femmagazine .com/students-call-for-ucla-administration-to-abolish-campus-police -presence/.

Chapter 6. (Beyond) Digital Literacy

1. Chanel M. Sutherland, "Digital Natives: 4 Ways Technology Has Changed 'the' Student," *Explorance*, March 31, 2016, https://explorance.com/blog/digital -natives-4-ways-technology-changed-student/. The term "digital natives" was originally coined in Marc Prensky, "Digital Natives, Digital Immigrants," *On the Horizon* 9, no. 5 (2001): 1–6, http://dx.doi.org/10.1108/10748120110424816.

2. Erika E. Smith, "The Digital Native Debate in Higher Education: A Comparative Analysis of Recent Literature," *Canadian Journal of Learning and Technology* 38, no. 3 (2012): 1–18.

3. Pauline Hope Cheong and Pratik Nyaupane, "Smart Campus Communication, Internet of Things, and Data Governance: Understanding Student Tensions and Imaginaries," *Big Data & Society* 9, no. 1 (2022): 1–13, https://doi .org/10.1177/20539517221092656.

4. Manuela Ekowo and Iris Palmer, "The Promise and Peril of Predictive Analytics in Higher Education: A Landscape Analysis," *New America*, October 24, 2016, https://www.newamerica.org/education-policy/policy -papers/promise-and-peril-predictive-analytics-higher-education/.

5. Electronic Privacy Information Center, "Student Privacy Bill of Rights," accessed May 21, 2023, https://epic.org/student-privacy-bill-of-rights/.

6. Luciana Pangrazio, "Reconceptualising Critical Digital Literacy," *Discourse: Studies in the Cultural Politics of Education* 37, no. 2 (2016): 163–174, http://dx.doi.org/10.1080/01596306.2014.942836.

7. Whitney Gegg-Harrison, "Against the Use of GPTZero and Other LLM-Detection Tools on Student Writing," *Medium*, February 27, 2016,

https://writerethink.medium.com/against-the-use-of-gptzero-and-other-llm-detection-tools-on-student-writing-b876b9d1b587/.

8. Karen Hao, "The Messy, Secretive Reality behind OpenAI's Bid to Save the World," *MIT Technology Review*, February 17, 2020, https://www.technologyreview.com/2020/02/17/844721/ai-openai-moonshot-elon-musk-sam-altman-greg-brockman-messy-secretive-reality/.

9. Keoni Mahelona, Gianna Leoni, Suzanne Duncan, and Miles Thompson, "OpenAI's Whisper Is Another Case Study in Colonialism," *PapaReo* (blog), January 24, 2023, https://blog.papareo.nz/whisper-is-another-case-study-in-colonisation/.

10. Chris Vallance, "Artificial Intelligence Could Lead to Extinction, Experts Warn," BBC News, May 30, 2023, https://www.bbc.com/news/uk-65746524.

11. Thanh Long Phan, "How ChatGPT Is Fine-Tuned Using Reinforcement Learning," *Dida*, April 11, 2023, https://dida.do/blog/chatgpt-reinforcement-learning.

12. For more on ChatGPT's inadequate data security, see Michael Kan, "OpenAI: Sorry, ChatGPT Bug Leaked Payment Info to Other Users," *PC Mag*, March 24, 2023, https://www.pcmag.com/news/openai-sorry-chatgpt-bug-leaked-payment-info-to-other-users/; For reporting on the working conditions of ChatGPT data annotators in Kenya, see Billy Perrigo, "Exclusive: OpenAI Used Kenyan Workers on Less Than $2 Per Hour to Make ChatGPT Less Toxic," *Time*, January 18, 2023, https://time.com/6247678/openai-chatgpt-kenya-workers/.

13. Khadijah Abdurahman (@UpFromTheCracks), Twitter Post, February 16, 2023, 12:29 p.m., https://mobile.twitter.com/UpFromTheCracks/status/1626272473950846983.

14. Chloe Xiang, "ChatGPT Can Replace the Underpaid Workers Who Train AI, Researchers Say," *Vice*, March 29, 2023, https://www.vice.com/en/article/ak3dwk/chatgpt-can-replace-the-underpaid-workers-who-train-ai-researchers-say/. For the original paper claiming that ChatGPT can label text more accurately than human annotation workers, see Fabrizio Gilardi, Meysam Alizadeh, and Mäel Kubli, "ChatGPT Outperforms Crowd-Workers for Text-Annotation Tasks," *arXiv* (2023): 1–13, https://arxiv.org/pdf/2303.15056.pdf.

15. Claire Chen, "AI Will Transform Teaching and Learning: Let's Get It Right," Stanford University, last modified March 9, 2023, https://hai.stanford.edu/news/ai-will-transform-teaching-and-learning-lets-get-it-right/.

16. Marika Cifor, Patricia Garcia, T. L. Cowan, Jasmine Rault, Tonia Sutherland, Anita Say Chan, Jennifer Rode, Anna Lauren Hoffmann, Niloufar Salehi, and Lisa Nakamura, *Feminist Data Manifest-No* (2019), https://www.manifestno.com/.

17. Autumm Caines, "Prior to (or instead of) Using ChatGPT with Your Students," *Autumn Caines* (blog), January 18, 2023, https://autumm.edtech.fm/2023/01/18/prior-to-or-instead-of-using-chatgpt-with-your-students/.

18. Dieuwertje Luitse and Wiebke Denkena, "The Great Transformer: Examining the Role of Large Language Models in the Political Economy of AI," *Big Data & Society* 8, no. 2 (July 2021): 10, https://doi.org/10.1177/20539517211047734.

19. Emily M. Bender, Timnit Gebru, Angelina McMillan-Major, and Smargaret Shmitchell, "On the Dangers of Stochastic Parrots: Can Language Models Be Too Big?," in *Proceedings of the ACM Conference on Fairness, Accountability, and Transparency* (2021): 610–623, https://doi.org/10.1145/3442188.3445922.

20. Civics of Technology, "Conduct EdTech Audit," accessed May 21, 2023, https://www.civicsoftechnology.org/edtechaudit.

21. Nassim Parvin and Anne Pollack, "Unintended by Design: On the Political Uses of 'Unintended Consequences,'" *Engaging Science, Technology, and Society* 6 (2020): 326, https://doi.org/10.17351/ests2020.497.

22. Neil Selwyn, *Education and Technology: Key Issues and Debates*, 2nd ed. (New York: Bloomsbury, 2016), 186.

23. CUNY Academic Commons, "We Help People at CUNY Connect, Learn, Teach, and More. What Will You Create?," accessed October 6, 2023, https://commons.gc.cuny.edu/.

24. Sarah Brown, "Professors, It's Time to 'Rate Your Campus Admin,'" *Chronicle of Higher Education*, September 2, 2022, https://www.proquest.com/docview/2720299094/.

25. Brian Clifton, Sam Lavignem, and Francis Tseng, "Predicting Financial Crime: Augmenting the Predictive Policing Arsenal," last modified April 25, 2017, https://arxiv.org/abs/1704.07826/.

26. Daniel Schiff, "Education for AI, *not* AI for Education: The Role of Education and Ethics in National AI Policy Strategies," *International Journal of Artificial Intelligence in Education* 32 (2022): 527–563, https://doi.org/10.1007/s40593-021-00270-2.

Conclusion. Against the Smart University

1. Ben Williamson, "The Hidden Architecture of Higher Education: Building a Big Data Infrastructure for the 'Smarter University,'" *International Journal of Higher Education* 15, no. 12 (2018): 4, https://doi.org/10.1186/s41239-018-0094-1.

2. Robert Ovetz, "The Algorithmic University: On-Line Education, Learning Management Systems, and the Struggle over Academic Labor," *Critical Sociology* 47, nos. 7/8 (2021): 1065–1084, https://doi.org/10.1177/0896920520948931.

3. AspirEDU, "Instructor Insight," accessed May 21, 2023, https://aspiredu.com/instructor-insight/.

4. Creatrix Campus, "Faculty Management System," accessed May 21, 2023, https://www.creatrixcampus.com/faculty-management-software.

5. Mary Clotilda Suvin, "Everything You Need to Know about an Effective Faculty Management Solution," Creatrix Campus, last modified December 12, 2022, https://www.creatrixcampus.com/blog/everything-you-need-know-about-effective-faculty-management-solution.

6. Juan Pablo Pardo-Guerra, *The Quantified Scholar: How Research Evaluations Transformed the British Social Sciences* (New York: Columbia University Press, 2022), 192.

7. Michael Kwet and Paul Prinsloo, "The 'Smart' Classroom: A New Frontier in the Age of the Smart University," *Teaching in Higher Education* 25, no. 4 (2020): 518, https://doi.org/10.1080/13562517.2020.1734922.

8. Roderick A. Ferguson, *The Reorder of Things: The University and Its Pedagogies of Minority Difference* (Minneapolis: University of Minnesota Press, 2012).

9. Susan S. Silbey, "How Not to Teach Ethics," *MIT Faculty Newsletter*, last modified September/October 2018, http://web.mit.edu/fnl/volume/311/silbey.html.

10. For more on Computer People for Peace, see Joan Greenbaum, "Questioning Tech Work," *AI Now Institute*, January 31, 2020, https://ainowinstitute.org/publication/questioning-tech-work-2. For more on the Polaroid Revolutionary Workers Movement, see Michael McCanne, "When Polaroid Workers Fought Apartheid," *Dissent*, August 14, 2020, https://www.dissentmagazine.org/online_articles/when-polaroid-workers-fought-apartheid/.

11. Sasha Costanza-Chock, *Design Justice: Community-Led Practices to Build the Worlds We Need* (Cambridge, MA: MIT Press, 2020), 211–217.

12. Suhauna Hussain, "'Google Has No Scruples': Employees Protest Google Cloud Conference over Israeli Military Contract," *Los Angeles Times*, August 29, 2023, https://www.latimes.com/business/story/2023-08-29/google-cloud-employees-protest-israeli-military-contract.

13. Costanza-Chock, *Design Justice*.

14. Our Data Bodies, "Who We Are," accessed May 21, 2023, https://www.odbproject.org/about-us-2/.

15. Analogue University, "Correlation in the Data University: Understanding and Challenging Targets-Based Performance-Management in Higher Education," *International Journal for Critical Geographies* 18, no. 6 (December 2019): 1184–1206.

16. Ian Gavigan and Jennifer Mittelstadt, "A New Deal for Eds and Meds," *Popular Resistance*, October 6, 2021, https://popularresistance.org/a-new-deal-for-eds-and-meds/.

17. Charlie Eaton, "Student Debt Cancellation on Campus," *Academe*, Fall 2022, https://www.aaup.org/article/%E2%80%8Bstudent-debt-cancellation-campus#.ZGkOS3bMKUk/.

18. Debt Collective, "Cancel Student Debt," accessed May 21, 2023, https://debtcollective.org/what-we-do/campaigns/student-debt/.

19. *People's Dispatch*, "US Supreme Court Strikes Down Student Debt Relief—Another Blow to Education for All," June 30, 2023, https://peoplesdispatch.org/2023/06/30/us-supreme-court-strikes-down-student-debt-relief-another-blow-to-education-for-all/.

20. Ian Gavigan and Jennifer Mittelstadt, "A New Deal for Eds and Meds," *Popular Resistance*, October 6, 2021, https://popularresistance.org/a-new-deal-for-eds-and-meds/.

21. Jennifer Mittelstadt, "Building a New Framework of Values for the University: Emerging from the Ivory Tower's Shadow; Interview with Davarian L. Baldwin," *Academe*, Fall 2022, https://www.aaup.org/article/building-new-framework-values-university#.ZGkOtnbMKUk/.

22. See Myya Helm, "Everyone at West Virginia University Knew Something Was Up: I Hate That They Were Right," *Slate*, August 18, 2023, https://slate.com/human-interest/2023/08/west-virginia-university-cuts-programs.html; and West Virginia University, "WVU Modernization Program," accessed October 3, 2023, https://strategicinitiatives.wvu.edu/wvu-modernization-program/about.

23. Abigail Boggs, Eli Meyerhoff, Nick Mitchell, and Zach Schwartz-Weinstein, "Abolitionist University Studies: An Invitation," *Abolition University*, https://abolition.university/invitation/#_ftn7.

Bibliography

Abdurahman, Khadijah (@UpFromTheCracks). Twitter post. February 16, 2023, 12:29 p.m. https://mobile.twitter.com/UpFromTheCracks/status/1626272473950846983.

Abrams, Richard M. "The US Military and Higher Education: A Brief History." *Annals of the American Academy* 502, no. 1 (1989): 15–28. https://doi.org/10.1177/0002716289502001002.

Adorno, Theodor, and Max Horkheimer. "The Culture Industry: Enlightenment as Mass Deception." In *Dialectic of Enlightenment*, 94–136. Stanford, CA: Stanford University Press, 2022.

Ahmed, Sara. *On Being Included: Racism and Diversity in Institutional Life*. Durham, NC: Duke University Press, 2012.

Ahmed, Sara. "Selfcare as Warfare." *Feministkilljoys* (blog), August 25, 2014. https://feministkilljoys.com/2014/08/25/selfcare-as-warfare/.

Albino, Vito, Umberto Berardi, and Rosa Maria Dangelico. "Smart Cities: Definitions, Dimensions, Performance, and Initiatives." *Journal of Urban Technology* 22, no.1 (2015): 3–21. https://doi.org/10.1080/10630732.2014.942092.

Alotaibi, Sultan Refa. "An Integrated Framework for Smart College Based on the Fourth Industrial Revolution." *International Transaction Journal of Engineering, Management, & Applied Sciences and Technologies* 12, no. 4 (2021): 1–18. https://tuengr.com/V12/12A4R.pdf.

Analogue University. "Correlation in the Data University: Understanding and Challenging Targets-Based Performance-Management in Higher Education." *International Journal for Critical Geographies* 18, no. 6 (December 2019): 1184–1206.

Andrejevic, Mark. "Estrangement 2.0." *World Picture* 6 (2011): 1–14.

Angwin, Julia, Jeff Larson, Surya Mattu, and Lauren Kirchner. "Machine Bias: There's Software Used across the County to Predict Future Criminals. And It's Biased against Blacks." *ProPublica*, May 23, 2016. https://www.propublica.org/article/machine-bias-risk-assessments-in-criminal-sentencing.

Anthology. "Blackboard Learn." Accessed October 19, 2023. https://www.anthology.com/products/teaching-and-learning/learning-effectiveness/blackboard-learn/.

Aoun, Joseph E. *Robot Proof: Higher Education in the Age of Artificial Intelligence*. Cambridge, MA: MIT Press, 2018.

Asif, Manan Ahmed. "Technologies of Power—from Area Studies to Data Sciences." *Spheres: Journal for Digital Cultures*, November 20, 2019. http://spheres-journal.org/technologies-of-power-from-area-studies-to-data-sciences/.

AspirEDU. "Instructor Insight." Accessed May 21, 2023. https://aspiredu.com/instructor-insight/.

Baker, Dominique J. "Pathways to Racial Equity in Higher Education: Modeling the Antecedents of State Affirmative Action Bans." *American Educational Research Journal* 56, no. 5 (2019): 1861–1895. https://doi.org/10.3102/0002831219833918.

Baldwin, Davarian L. *In the Shadow of the Ivory Tower: How Universities Are Plundering Our Cities*. New York: Bold Type Books, 2021.

Balzacq, Thierry. Preface to *Securitization Theory: How Security Problems Emerge and Dissolve*. Edited by Thierry Balzacq, xiii–xiv. London: Routledge, 2011.

Barshay, Jill, and Sasha Aslanian. "Under a Watchful Eye." *AMP Reports*, August 9, 2019. https://www.apmreports.org/episode/2019/08/06/college-data-tracking-students-graduation.

Beer, David. *Metric Power*. London: Palgrave Macmillan, 2016.

Benanav, Aaron. *Automation and the Future of Work*. London: Verso Books, 2022.

Bender, Emily M., Timnit Gebru, Angelina McMillan-Major, and Smargaret Shmitchell. "On the Dangers of Stochastic Parrots: Can Language Models Be Too Big?" In *Proceedings of the ACM Conference on Fairness, Accountability, and Transparency* (2021): 610–623. https://doi.org/10.1145/3442188.3445922.

Benjamin, Ruha. *Race after Technology: Abolitionist Tools for the New Jim Code*. Cambridge: Polity, 2019.

Biddle, Sam. "Amazon Admits Giving Ring Camera Footage to Police without a Warrant or Consent." *Intercept*, July 13, 2020. https://theintercept.com/2022/07/13/amazon-ring-camera-footage-police-ed-markey/.

Blackboard. "Enhance Your Student Success Strategy." Last modified 2017. https://www.blackboard.com/resources/enhance-your-student-success-strategy/.

Blas, Zach, and Jacob Gaboury. "Biometrics and Opacity: A Conversation." *Camera Obscura* 31, no. 2 (2016): 155–165.

Boggs, Abigail. "On Borders and Academic Freedom: Noncitizen Students and the Limits of Rights." *AAUP Journal of Academic Freedom* 11 (2020): 1–17.

Boggs, Abigail, Eli Meyerhoff, Nick Mitchell, and Zach Schwartz-Weinstein. "Abolitionist University Studies: An Invitation." *Abolition University*. https://abolition.university/invitation/#_ftn7.

Boggs, Abigail, and Nick Mitchell. "Critical University Studies and the Crisis Consensus." *Feminist Studies* 44, no. 2 (2018): 432–463. https://doi.org/10.15767/feministstudies.44.2.0432.

Bousquet, Marc. *How the University Works: Higher Education and the Low-Wage Nation*. New York: New York University Press, 2008.

Bowman, Emma. "Here's What Happened when Affirmative Action Ended at California Public Colleges." NPR, June 30, 2023. https://www.npr.org/2023/06/30/1185226895/heres-what-happened-when-affirmative-action-ended-at-california-public-colleges.

Brief for the United States as Amicus Curiae Supporting Respondents, Students for Fair Admissions v. University of North Carolina et al., 600 U.S. _____ (2023) (No. 21–707), https://www.justice.gov/crt/case-document/file/1523991/download.

Broussard, Meredith. *Artificial Unintelligence: How Computers Misunderstand the World*. Cambridge, MA: MIT Press, 2018.

Brown, Dalvin. "Alexa Goes to College: Echo Dots Move into Dorms on Campus." *USA Today*, September 6, 2018. https://www.usatoday.com/story/money/2018/09/06/college-students-echo-dots-dorm-rooms/1087251002/.

Brown, Sarah. "Professors, It's Time to 'Rate Your Campus Admin.'" *Chronicle of Higher Education*, September 2, 2022. https://www.proquest.com/docview/2720299094/.

Browne, Simone. *Dark Matters: On the Surveillance of Blackness*. Durham, NC: Duke University Press, 2015.

Brunila, Kristiina, and Leena-Maija Rossi. "Identity Politics, the Ethos of Vulnerability, and Education." *Educational Philosophy and Theory* 50, no. 3 (2017): 287–298. https://doi.org/10.1080/00131857.2017.1343115.

Bualamwini, Joy, and Timnit Gebru. "Gender Shades: Intersectional Accuracy Disparities in Commercial Gender Classification." In *Proceedings of the First Conference on Fairness, Accountability and Transparency (Proceedings of Machine Learning Research)* 81 (2018): 77–91. http://proceedings.mlr.press/v81/buolamwini18a.html.

Byrd, W. Carson. *Behind the Diversity Numbers: Achieving Racial Equity on Campus*. Boston: Harvard Education Press, 2021.

Caines, Autumn. "Prior to (or instead of) Using ChatGPT with Your Students." *Autumn Caines* (blog), January 18, 2023. https://autumm.edtech.fm/2023/01/18/prior-to-or-instead-of-using-chatgpt-with-your-students/.

Cameron, Andy, and Richard Barbrook. "The Californian Ideology." *Science as Culture* 6, no. 1 (1996): 44–72.

CampusLogic. "Higher Education Financial Aid Software." Accessed May 21, 2023. https://campuslogic.com/.

Capture. "Make Every Interaction Count." Accessed May 21, 2023. https://www.capturehighered.com/.

Capture. "Marketing & Recruitment Solutions." Accessed June 20, 2022. https://www.capturehighered.com/solutions/.

Capture. "The 1-2-3 of Machine Learning to Power Your Predictive Model." Accessed August 24, 2021. https://www.youtube.com/watch?v=5p5cYV5AtB0.

Capture. "Delivering the Industry's Highest-Qualified Student Inquiries." Accessed May 21, 2023. https://www.capturehighered.com/wp-content/uploads/2021/09/Capture-Inquiries-Sales.pdf.

Cerci, Sena. "Embodying Self-Tracking: A Feminist Exploration of Collective Meaning-Making of Self-Tracking Data." MA, Malmo University. 2018.

Chang, Ho-Chun Herbert, Matt Bui, and Charlton McIlwain. "Targeted Ads and/as Racial Discrimination: Exploring Trends in New York City Ads for College Scholarships." *arXiv* (2019). https://arxiv.org/pdf/2109.15294.pdf.

Chatterjee, Piya, and Sunaina Maira. "The Imperial University: Race, War, and the Nation-State." In *The Imperial University: Academic Repression and Scholarly Dissent*, edited by Piya Chatterjee and Sunaina Maira, 1–50. Minneapolis: University of Minnesota Press, 2014.

Chen, Claire. "AI Will Transform Teaching and Learning: Let's Get It Right." Stanford University. Last modified March 9, 2023. https://hai.stanford.edu/news/ai-will-transform-teaching-and-learning-lets-get-it-right/.

Cheong, Pauline Hope, and Pratik Nyaupane. "Smart Campus Communication, Internet of Things, and Data Governance: Understanding Student Tensions and Imaginaries." *Big Data & Society* 9, no. 1 (2022): 1–13. https://doi.org/10.1177/20539517221092656.

Chin, Monica. "College Student Sues Proctorio after Source Code Copyright Claim." *Verge*, May 5, 2021. https://www.theverge.com/2021/4/22/22397499/proctorio-lawsuit-electronic-frontier-foundation-test-proctoring-software.

Cifor, Marika, Patricia Garcia, T. L. Cowan, Jasmine Rault, Tonia Sutherland, Anita Say Chan, Jennifer Rode, Anna Lauren Hoffmann, Niloufar Salehi, and Lisa Nakamura. *Feminist Data Manifest-No* (2019). https://www.manifestno.com/.

Civics of Technology. "Conduct EdTech Audit." Accessed May 21, 2023. https://www.civicsoftechnology.org/edtechaudit.

Clifton, Brian, Sam Lavignem, and Francis Tseng. "Predicting Financial Crime: Augmenting the Predictive Policing Arsenal." *arXiv* (April 25, 2017). https://arxiv.org/abs/1704.07826/.

Coalition for Critical Technology. "Abolish the #TechtoPrisonPipeline." *Medium*, June 23, 2020. https://medium.com/@CoalitionForCriticalTechnology/abolish-the-techtoprisonpipeline-9b5b14366b16.

Cohen, Sol. "The Mental Hygiene Movement, the Development of Personality and the School: The Medicalization of American Education." *History of Education Quarterly* 23, no. 2 (1983): 123–149. https://doi.org/10.2307/368156.

Coherent Market Insights. "Emotion Detection and Recognition Market to Surpass US$ 40.5 Billion by 2030." *Global Newswire*, March 27, 2023. https://www.globenewswire.com/news-release/2023/03/27/2634806/0/en/Emotion-Detection-and-Recognition-Market-to-surpass-US-40-5-Billion-by-2030-Says-Coherent-Market-Insights-CMI.html.

Cole, Eddie R. "The Racist Roots of Campus Policing." *Washington Post*, June 2, 2021. https://www.washingtonpost.com/outlook/2021/06/02/racist-roots-campus-policing/.

Collabco. "From Digital Student to Smart Citizen." White paper. Accessed October 10, 2023. https://myday.collabco.com/wp-content/uploads/from-digital-student-to-smart-citizen-whitepaper.pdf.

Comeaux, Eddie. "Stereotypes, Control, Hyper-surveillance, and Disposability of the NCAA Division I Black Male Athletes." *New Directions for Student Services*, no. 163 (2018): 33–42. https://doi.org/10.1002/ss.20268.

Cooky, Cheryl, Jasmine R. Lindabary, and Danielle J. Corple. "Navigating Big Data Dilemmas: Feminist Holistic Reflexivity in Social Media Research." *Big Data & Society* 5, no. 2 (2018): 1–12. https://doi.org/10.1177/2053951718807731.

Costanza-Chock, Sasha. *Design Justice: Community-Led Practices to Build the Worlds We Need.* Cambridge, MA: MIT Press, 2020.

Costanza-Chock, Sasha. "Design Justice: Towards an Intersectional Feminist Framework for Design Theory and Practice." In *Proceedings of the Design Research Society* (2018): 1–14. https://ssrn.com/abstract=3189696.

Cottom, Tressie McMillan. *Lower Ed: The Troubling Rise of For-Profit Colleges in the New Economy.* New York: New Press, 2017.

Coy, Peter. "The College Admissions Scandal Presses Our 'Unfairness' Button." *Bloomberg*, March 14, 2019. https://www.bloomberg.com/news/articles/2019-03-14/the-college-admissions-scandal-presses-our-unfairness-button.

Crary, Jonathan. *24/7: Late Capitalism and the Ends of Sleep.* London: Verso, 2014.

Creatrix Campus. "Faculty Management System." Accessed May 21, 2023. https://www.creatrixcampus.com/faculty-management-software.

Crespo, Monica. "Fair, Transparent, and Accountable Algorithmic Decision-Making: What Is the Role of Human-in-the-Loop?" *iSChannel* 17, no. 1 (2022): 28–41. https://ischannel.lse.ac.uk/articles/208/.

Crow, Michael M., and William B. Dabars. *Designing the New American University*. Baltimore: Johns Hopkins University Press, 2015.

CUNY Academic Commons. "We Help People at CUNY Connect, Learn, Teach, and More. What Will You Create?" Accessed October 6, 2023. https://commons.gc.cuny.edu/.

CyberPsyc Software Solutions. "Privacy Policy." Accessed January 31, 2019. https://app.welltrack.com/privacy-policy.

Daniels, Mitch. "Someone Is Watching You." *Washington Post*, March 27, 2018. https://www.washingtonpost.com/opinions/its-okay-to-be-paranoid-someone-is-watching-you/2018/03/27/1a161d4c-2327-11e8-86f6-54bfff693d2b_story.html/.

Dart, Tom. "University of Texas: Eco-conscious Campus and Major Fracking Landlord." *Guardian*, October 10, 2018. https://www.theguardian.com/us-news/2018/oct/10/university-of-texas-eco-conscious-campus-and-major-fracking-landlord.

Davenport, Thomas H., and John C. Beck. *The Attention Economy: Understanding the New Currency of Business*. Cambridge, MA: Harvard Business Review Press, 2002.

Davis, Angela. *Are Prisons Obsolete?* New York: Seven Stories Press, 2023.

Debt Collective. "Cancel Student Debt." Accessed May 21, 2023. https://debtcollective.org/what-we-do/campaigns/student-debt/.

Debt Collective. "End Student Debt!" Accessed January 25, 2021. https://strike.debtcollective.org/.

DeGeurin, Mark. "The College Board Is Licensing the Personal Data of Students Taking the SAT to Colleges So They Can Reject More Students and Inflate Admissions Numbers." *Insider*, November 6, 2019. https://www.insider.com/college-board-sat-student-data-colleges-to-reject-students-admissions-2019-11.

Delavega, Elena and Gregory M. Blumenthal. "2023 Memphis Poverty Fact Sheet." Accessed October 20, 2023. https://www.memphis.edu/socialwork/research/2023-poverty-fact-sheet-final.pdf.

Deloitte. "Smart Campus: the Next-Generation Connected Campus." Last modified 2019. https://www2.deloitte.com/content/dam/Deloitte/us/Documents/strategy/the-next-generation-connected-campus-deloitte.pdf.

Dent, Steve. "Colorado College Students Were Secretly Used to Train Facial Recognition." *Engadget*, May 28, 2019. https://www.engadget.com/2019/05/28/uccs-facial-recognition-study-students/?guccounter=1.

Department of Justice and Department of Education. "Questions and Answers Regarding the Supreme Court Decision in *Students for Fair Admissions, Inc. v. Harvard College and University of North Carolina*."

August 14, 2023. https://www.justice.gov/d9/2023-08/post-sffa_resource_faq_final_508.pdf.

Dickerson, Caitlin. "'Demeaned and Humiliated': What Happened to These Iranians at U.S. Airports." *New York Times*, January 27, 2020. https://www.nytimes.com/2020/01/25/us/iran-students-deported-border.html.

D'Ignazio, Catherine, and Lauren F. Klein. *Data Feminism*. Cambridge, MA: MIT Press, 2023.

Diaz-Strong, Daysi, Christina Gómez, María E. Luna-Duarte, Erica R. Meiners, and Luvia Valentin. "Organizing Tensions—from the Prison to the Military-Industrial-Complex." *Social Justice* 36, no. 2 (2009/2010): 73–84. https://www.jstor.org/stable/29768538.

Divoky, Diane. "Cumulative Records: Assault on Privacy." *Learning* 2, no. 18 (1973): 18–21.

Dixon-Román, Ezekiel J. *Inheriting Possibility: Social Reproduction and Quantification in Education*. Minneapolis: University of Minnesota Press, 2017.

Dobbe, Roel, Sarah Dean, Thomas Gilbert, and Nitin Kohli. "A Broader View on Bias in Automated Decision-Making: Reflecting on Epistemology and Dynamics." *Proceedings of ICML Workshop on Fairness, Accountability, and Transparency in Machine Learning* (2018): 1–5. https://arxiv.org/pdf/1807.00553.pdf.

Dong, Zhao Yang, Yuchen Zhang, Christine Yip, Sharon Swift, and Kim Beswick. "Smart Campus: Definition, Framework, Technologies, and Services." *Institution of Engineering and Technology* (2020): 1–12. https://doi.org/10.1049/iet-smc.2019.0072.

Doty, Philip. "Library Analytics as Moral Dilemmas for Academic Librarians." *Journal of Academic Librarianship* 46, no. 4 (July 2020): 1–5. https://doi.org/10.1016/j.acalib.2020.102141.

Doyle, Jennifer. *Campus Sex, Campus Security*. Cambridge: Semiotext(e), 2015.

Dubrofsky, Rachel, and Shoshana Amielle Magnet. Introduction to *Feminist Surveillance Studies*, edited by Rachel Dubrofsky and Shoshana Amielle Magnet, 1–18. Durham, NC: Duke University Press, 2015.

Duffy, Brooke Erin, and Ngai Keung Chan. "'You Never Really Know Who's Looking': Imagined Surveillance across Social Media Platforms." *New Media & Society*, no. 1, (January 2019): 119–138. https://doi.org/10.1177/1461444818791318.

Dunbar, Michael S., Lisa Sontag-Padilla, Rajeev Ramchand, Rachana Seelam, and Bradley D. Stein. "Mental Health Service Utilization among Lesbian, Gay, Bisexual, and Questioning or Queer College Students." *Journal of Adolescent Health* 61, no. 3 (2017): 294–301. https://doi.org/10.1016/j.jadohealth.2017.03.008.

Eaton, Charlie. "Student Debt Cancellation on Campus." *Academe*, Fall 2022. https://www.aaup.org/article/%E2%80%8Bstudent-debt-cancellation-campus#.ZGkOS3bMKUk/.

EDUCAUSE. Accessed May 20, 2023, https://www.educause.edu/.

EDUCAUSE. "CAUSE History." Accessed October 17, 2023. https://www.educause.edu/about/mission-and-organization/our-history/cause-history.

EDUCAUSE. "Higher Education Community Vendor Assessment Toolkit." Last modified December 17, 2021. https://library.educause.edu/resources/2020/4/higher-education-community-vendor-assessment-toolkit.

EduNav. "The EduNav Suite." Accessed October 1, 2023. https://edunav.com/.

EduNav. "EduNav SmartPlan." Accessed October 1, 2023, https://edunav.com/edunav-smartplan/.

Ekowo, Manuela, and Iris Palmer. "The Promise and Peril of Predictive Analytics in Higher Education: A Landscape Analysis." *New America*, October 24, 2016. https://www.newamerica.org/education-policy/policy-papers/promise-and-peril-predictive-analytics-higher-education/.

Electronic Privacy Information Center. "Student Privacy Bill of Rights." Accessed May 21, 2023. https://epic.org/student-privacy-bill-of-rights/.

Ellucian (blog). "Technology Fosters Deeper Campus Connections." Accessed May 19, 2023. https://www.ellucian.com/blog/technology-fosters-deeper-campus-connections.

Eubanks, Virginia. "Want to Predict the Future of Surveillance? Ask Poor Communities." *American Prospect*, January 15, 2014. https://prospect.org/power/want-predict-future-surveillance-ask-poor-communities./.

Evans, Gabe, Nick Mitchell, and Taylor Wondergem. "Scenes from the Wildcat Strike: A Documentary History." *Critical Ethnic Studies* 6, no. 2 (2021). https://doi.org/10.5749/CES.0602.wildcat-strike.

Evans, Teresa M., Lindsay Bira, Jazmin Beltran Gastelum, L Todd Weiss, and Nathan L Vanderford. "Evidence for a Mental Health Crisis in Graduate Education." *Nature Biotechnology* 36 (2018): 282–284.

Felt, Ulrike, Rayvon Fouché, Clark A. Miller, and Laurel Smith-Doerr, eds. *The Handbook of Science and Technology Studies*. 4th ed. Cambridge, MA: MIT Press, 2016.

Ferguson, Andrew Guthrie. "Policing Predictive Policing." *Washington University Law Review* 94, no. 5 (2017): 1115–1194. https://papers.ssrn.com/sol3/papers.cfm?abstract_id=2765525.

Ferguson, Roderick A. *The Reorder of Things: The University and Its Pedagogies of Minority Difference*. Minneapolis: University of Minnesota Press, 2012.

Ferraro, David. "Psychology in the Age of Austerity." *Psychotherapy and Politics International* 14, no. 1 (2016): 17–24. https://doi.org/10.1002/ppi.1369.

Fiebig, Tobias, Seda Gürses, Carlos H. Gañán, Erna Kotkamp, Fernando Kuipers, Martina Lindorfer, Menghua Prisse, and Taritha Sari. "Heads in the Clouds: Measuring the Implications of Universities Migrating to Public Clouds." *arXiv* (2021): 1–34. https://arxiv.org/abs/2104.09462/.

Fight for the Future and Students for Sensible Drug Policy. "Stop Facial Recognition on Campus." Accessed May 19, 2023. https://www.banfacialrecognition.com/campus/.

Foucault, Michel. *The Care of the Self: The History of Sexuality*. Vol. 3. New York: Pantheon, 1986.

Fourcade, Marion and Kieran Healy. "Classification Situations: Life-Chances in the Neoliberal Era." *Accounting, Organizations and Society* 38, no. 8 (November 2013): 559–572. https://doi.org/10.1016/j.aos.2013.11.002.

Francis, Peter, Christine Broughan, Carly Foster, and Caroline Wilson. "Thinking Critically about Learning Analytics, Student Outcomes, and Equity of Attainment." *Assessment & Evaluation in Higher Education* 45, no. 6 (December 2019): 811–821. https://doi.org/10.1080/02602938.2019.1691975.

Fraser, Keith. "Software Company Sues UBC Employee over Tweets Involving Confidential Videos." *Province*, September 3, 2020. https://theprovince.com/news/software-company-sues-ubc-employee-over-tweets-involving-confidential-videos/wcm/551266ea-a842-4a48-a8e1-1f3fb732f911/.

Fusell, Sidney. "Did a University Use Facial Recognition to ID Student Protestors?" *Wired*, November 18, 2020. https://www.wired.com/story/did-university-use-facial-recognition-id-student-protesters/.

Fusell, Sidney. "You No Longer Own Your Face." *Atlantic*, June 27, 2019. https://www.theatlantic.com/technology/archive/2019/06/universities-record-students-campuses-research/592537/.

Gabbidon, Shaun L. "Racial Profiling by Store Clerks and Security Personnel in Retail Establishments: An Exploration of 'Shopping While Black.'" *Journal of Contemporary Criminal Justice* 19, no. 3 (2003): 345–364. https://doi.org/10.1177/1043986203254531.

Gaffney, Christopher, and Cerianne Robertson. "Smarter Than Smart: Rio de Janeiro's Flawed Emergence as a Smart City." *Journal of Urban Technology* 25, no. 3 (April 2016): 47–64. https://doi.org/10.1080/10630732.2015.1102423.

Galligan, Claire, Hannah Rosenfeld, Molly Kleinman, and Shobita Parthasarathy. "Cameras in the Classroom: Facial Recognition Technology in Schools." *Technology Assessment Project Report* (2020): 1–115. https://stpp.fordschool.umich.edu/sites/stpp/files/uploads/file-assets/cameras_in_the_classroom_full_report.pdf.

Gandy, Oscar H. *The Panoptic Sort: A Political Economy of Personal Information*. Boulder, CO: Westview, 1993.

Gardner, Lee. "Students under Surveillance? Data-Tracking Enters a Provocative New Phase." *Chronicle of Higher Education*, October 13, 2019. https://www.chronicle.com/article/students-under-surveillance/.

Gates, Bill. "Prepared Remarks at the National Conference of State Legislatures." Bill & Melinda Gates Foundation. Accessed October 9, 2023. https://www.gatesfoundation.org/ideas/speeches/2009/07/bill-gates-national-conference-of-state-legislatures-ncsl.

Gates, Bill. *The Road Ahead*. London: Penguin, 1996.

Gavigan, Ian, and Jennifer Mittelstadt. "A New Deal for Eds and Meds." *Popular Resistance*, October 6, 2021. https://popularresistance.org/a-new-deal-for-eds-and-meds/.

Gee, Georgia. "Un-alarmed: AI Tries (and Fails) to Detect Weapons in Schools." *Intercept*, May 7, 2023. https://theintercept.com/2023/05/07/ai-gun-weapons-detection-schools-evolv/.

Gegg-Harrison, Whitney. "Against the Use of GPTZero and Other LLM-Detection Tools on Student Writing." *Medium*, February 27, 2016. https://writerethink.medium.com/against-the-use-of-gptzero-and-other-llm-detection-tools-on-student-writing-b876b9d1b587/.

Ghaffary, Shirin. "Amazon Fired Chris Smalls: Now the New Union Leader Is One of Its Biggest Problems." *Vox*, June 7, 2022. https://www.vox.com/recode/23145265/amazon-fired-chris-smalls-union-leader-alu-jeff-bezos-bernie-sanders-aoc-labor-movement-biden.

Gidaris, Constantine. "Surveillance Capitalism, Datafication, and Unwaged Labour: The Rise of Wearable Fitness Devices and Interactive Life Insurance." *Surveillance & Society* 17, nos. 1/2 (2019): 132–138. https://doi.org/10.24908/ss.v17i1/2.12913.

Gilardi, Fabrizio, Meysam Alizadeh, and Mäel Kubli. "ChatGPT Outperforms Crowd-Workers for Text-Annotation Tasks." *arXiv* (2023): 1–13. https://arxiv.org/pdf/2303.15056.pdf.

Gill, Rosalind, and Ngaire Donaghue. "Resilience, Apps and Reluctant Individualism: Technologies of Self in the Neoliberal Academy." *Women's Studies International Forum* 54 (January/February 2016): 91–99. https://doi.org/10.1016/j.wsif.2015.06.016.

Gillespie, Tarleton. "The Relevance of Algorithms." In *Media Technologies*, edited by Tarleton Gillespie, Pablo Boczkowshi, and Kirsten Foot, 167–194. Cambridge, MA: MIT Press, 2012.

Gilliard, Chris. "How Ed Tech Is Exploiting Students." *Chronicle of Higher Education*, April 8, 2018. https://www.chronicle.com/article/how-ed-tech-is-exploiting-students/.

Gilliard, Chris. "Pedagogy and the Logic of Platforms." In *Open at the Margins: Critical Perspectives on Open Education*, edited by Maja Bali, Catherine

Cronin, Laura Czerniewicz, Robin DeRosa, and Rajiv Jhangiani, 115–118. Montreal: Rebus Community, 2020.

Gilliom, John. *Overseers of the Poor: Surveillance, Resistance, and the Limits of Privacy*. Chicago: University of Chicago Press, 2001.

Ginsberg, Olivia. "UM Counseling Center offers New App to Promote Mental Health." *Miami Hurricane*, November 5, 2018. https://www.themiamihurricane.com/2018/11/05/um-counseling-center-offers-new-app-to-promote-mental-health/.

Gitelman, Lisa, and Virginia Jackson. Introduction to *Raw Data Is an Oxymoron*, edited by Lisa Gitelman, 1–14. Cambridge, MA: MIT Press, 2013.

Godrej, Farah. "Neoliberalism, Militarization, and the Price of Dissent: Policing Protest at the University of California." In *The Imperial University: Academic Repression and Scholarly Dissent*, edited by Piya Chatterjee and Sunaina Maira, 125–143. Minneapolis: University of Minnesota Press, 2014.

Goff, Jay W., and Christopher M. Shaffer. "Big Data's Impact on College Admission Practices and Recruitment Strategies." In *Building a Smarter University: Big Data, Innovation, and Analytics*, edited by Jason Lane, 93–120. Albany: State University of New York Press, 2014.

Graham, Rebecca Dolinsky, and Amanda Konradi. "Contextualizing the 1990 Campus Security Act and Campus Sexual Assault in Intersectional and Historical Terms." *Journal of Aggression, Conflict and Peace Research* 10, no. 2 (2018): 93–102.

Green, Adrienne. "The Cost of Balancing Academia and Racism." *Atlantic*, January 21, 2016. https://www.theatlantic.com/education/archive/2016/01/balancing-academia-racism/424887/.

Green, Ben. "Data Science as Political Action." *Journal of Social Computing* 2, no. 3 (September 2021): 249–265. https://doi.org/10.23919/JSC.2021.0029.

Green, Ben. "'Fair' Risk Assessments: A Precarious Approach for Criminal Justice Reform." *5th Workshop on Fairness, Accountability, and Transparency in Machine Learning* (2018): 1–5. https://scholar.harvard.edu/files/bgreen/files/18-fatml.pdf.

Green, Ben. "'Good' Isn't Good Enough." *AI for Social Good Workshop at NeurIPS* (2019): 1–7. https://www.benzevgreen.com/wp-content/uploads/2019/11/19-ai4sg.pdf.

Green, Ben, and Lily Hu. "The Myth in the Methodology: Towards a Recontextualization of Fairness in Machine Learning." *Machine Learning: The Debates Workshop, 35th International Conference on Machine Learning* (2018): 1–5. https://econcs.seas.harvard.edu/files/econcs/files/green_icml18.pdf.

Greenbaum, Joan. "Questioning Tech Work." *AI Now Institute*, January 31, 2020. https://ainowinstitute.org/publication/questioning-tech-work-2.

Greenfield, Adam. *Against the Smart City*. New York: Do Projects, 2013.

Haggerty, Kevin D., and Richard V. Ericson, "The Surveillant Assemblage." *British Journal of Sociology* 51, no. 4 (2000): 605–622.

Hahn, Alicia. "2023 Student Loan Debt Statistics: Average Student Loan Debt." *Forbes*, last modified July 16, 2023. https://www.forbes.com/advisor/student-loans/average-student-loan-debt-statistics/.

Halberstadt, Amy G., Vanessa L. Castro, Qiao Chu, Fantasy T. Lozada, and Calvin M. Sims. "Preservice Teachers' Racialized Emotion Recognition, Anger Bias, and Hostility Attributions." *Contemporary Educational Psychology* 54 (July 2018): 125–138. https://doi.org/10.1016/j.cedpsych.2018.06.004.

Haldane, Andrew G. "Ideas and Institutions—a Growth Story." Speech to the Guild Society, University of Oxford, May 23, 2018. https://www.bis.org/review/r180627e.pdf.

Halpern, Orit, and Robert Mitchell. *The Smartness Mandate*. Cambridge, MA: MIT Press, 2022.

Hanna, Alex, Emily Denton, Andrew Smart, and Jamilia Smith-Loud. "Towards a Critical Race Methodology in Algorithmic Fairness." *Proceedings of the 2020 Conference on Fairness, Accountability, and Transparency* (2020): 501–512. https://doi.org/10.48550/arXiv.1912.03593.

Hansen, Morten, and Janja Komljenovic. "Automating Learning Situations in EdTech: Techno-Commercial Logic of Assetisation." *Postdigital Science and Education* 5, no. 1 (2022): 100–116. https://doi.org/10.1007/s42438-022-00359-4.

Hao, Karen. "The Messy, Secretive Reality Behind OpenAI's Bid to Save the World." *MIT Technology Review*, February 17, 2020. https://www.technologyreview.com/2020/02/17/844721/ai-openai-moonshot-elon-musk-sam-altman-greg-brockman-messy-secretive-reality/.

Haraway, Donna. "Situated Knowledges: The Science Question in Feminism and the Privilege of Partial Perspective." *Feminist Studies* 14, no. 3 (1988): 575–599. https://doi.org/10.2307/3178066.

Harvey, Adam, and Jules LaPlace. "UnConstrained College Students Dataset." Accessed February 4, 2024. https://exposing.ai/uccs/.

Harwell, Drew. "Colleges are Turning Students' Phones into Surveillance Machines, Tracking the Locations of Hundreds of Thousands." *Washington Post*, December 24, 2019. https://www.washingtonpost.com/technology/2019/12/24/colleges-are-turning-students-phones-into-surveillance-machines-tracking-locations-hundreds-thousands/.

Hawkins, Andrew J. "Alphabet's Sidewalk Labs Shuts Down Toronto Smart City Project." *Verge*, May 7, 2020. https://www.theverge.com/2020/5/7/21250594/alphabet-sidewalk-labs-toronto-quayside-shutting-down.

Helhoski, Anna. "Why the Student Debt Crisis Hits Black Borrowers Harder." *Nerd Wallet*, last modified February 7, 2018. https://www.nerdwallet.com/blog/loans/student-loans/black-student-debt-crisis/.

Helm, Myya. "Everyone at West Virginia University Knew Something Was Up: I Hate That They Were Right." *Slate*, August 18, 2023. https://slate.com/human-interest/2023/08/west-virginia-university-cuts-programs.html.

Hirsch, Laura, and Michelle Castillo. "Amazon Has Big Plans for Alexa Ads in 2018: It's Discussing Options with P&G, Clorox, and Others." CNBC, January 2, 2018. https://www.cnbc.com/2018/01/02/amazon-alexa-is-opening-up-to-more-sponsored-product-ads.html.

Hollister, Sean. "Today I Learned Amazon Has a Form So Police Can Get My Data without Permission or a Warrant." *Verge*, July 15, 2022. https://www.theverge.com/2022/7/14/23219419/amazon-ring-law-enforcement-no-warrant-no-consent.

Hong, Sun-Ha. "Predictions without Futures." *History and Theory* 61, no. 3 (2022): 371–390. https://doi.org/10.1111/hith.12269.

Hussain, Suhauna. "'Google Has No Scruples': Employees Protest Google Cloud Conference over Israeli Military Contract." *Los Angeles Times*, August 29, 2023. https://www.latimes.com/business/story/2023-08-29/google-cloud-employees-protest-israeli-military-contract.

Hutchinson, Les, and Maria Novotny. "Teaching a Critical Literacy of Wearables: A Feminist Surveillance as Care Pedagogy." *Computers and Composition* 50 (December 2018): 105–120. https://doi.org/10.1016/j.compcom.2018.07.006.

Jansson, Åsa. "From Self-Help to CBT: Regulating Emotion in a (Neo)liberal World." *The History of Emotions Blog*, December 11, 2017. https://emotionsblog.history.qmul.ac.uk/2017/12/from-self-help-to-cbt-regulating-emotion-in-a-neoliberal-world/.

Jasanoff, Sheila, and Sang-Hyun Kim. *Dreamscapes of Modernity: Sociotechnical Imaginaries and the Fabrication of power*. Chicago: University of Chicago Press, 2015.

Jones, Martin. "What Is a Smart Campus and the Benefits to College Students and Faculty." *Cox Business*. Accessed May 21, 2023. https://www.coxblue.com/what-is-a-smart-campus-and-the-benefits-to-college-students-and-faculty/.

Kahn, Jeffery. "Ronald Reagan Launched Political Career Using the Berkeley Campus as a Target." *UC Berkeley News*, June 8, 2004. https://www.berkeley.edu/news/media/releases/2004/06/08_reagan.shtml.

Kan, Michael. "OpenAI: Sorry, ChatGPT Bug Leaked Payment Info to Other Users." *PC Mag*, March 24, 2023. https://www.pcmag.com/news/openai-sorry-chatgpt-bug-leaked-payment-info-to-other-users/.

Karlis, Nicole. "Emails Show UC Santa Cruz Police Used Military Surveillance to Suppress Student Strike." *Salon*, May 18, 2020. https://www.salon.com/2020/05/18/emails-show-uc-santa-cruz-police-used-military-surveillance-to-suppress-grad-student-strike/.

Kazar, Adrianna, and Daniel Maxey. "The Changing Faculty and Student Success: Selected Research on Connections between Non-tenure-track Faculty and Student Learning." Pulias Center for Higher Education (2012). https://files.eric.ed.gov/fulltext/ED532273.pdf.

Keyes, Os. "The Misgendering Machines: Trans/HCI Implications of Automatic Gender Recognition." *Proceedings of the ACM on Human-Computer Interaction* 2 (2018): 1–22. https://doi.org/10.1145/3274357.

Keyes, Os, Nikki Stevens, and Jacqueline Wernimont. "The Government Is Using the Most Vulnerable People to Test Facial Recognition Software." *Slate*, March 17, 2019. https://slate.com/technology/2019/03/facial-recognition-nist-verification-testing-data-sets-children-immigrants-consent.html.

Knox, Jeremy, Ben Williamson, and Sian Bayne. "Machine Behaviourism: Future Visions of 'Learnification' and 'Datafication' across Human and Digital Technologies." *Learning, Media and Technology* 45, no. 1 (April 29, 2020): 31–45. https://doi.org/10.1080/17439884.2019.1623251.

Komljenovic, Janja. "The Rise of Education Rentiers: Digital Platforms, Digital Data and Rents." *Learning, Media and Technology* 46, no. 3 (June 2020): 230–332. https://doi.org/10.1080/17439884.2021.1891422.

Kraft, David P. "One Hundred Years of College Mental Health." *Journal of American College Health* 59, no. 6 (2011): 477–481.

Kundnani, Arun, and Deepa Kumar. "Race, Surveillance, and Empire." *International Socialist Review*, no. 96 (2016). https://isreview.org/issue/96/race-surveillance-and-empire.

Kwet, Michael, and Paul Prinsloo. "The 'Smart' Classroom: A New Frontier in the Age of the Smart University." *Teaching in Higher Education* 25, no. 4 (2020): 510–526. https://doi.org/10.1080/13562517.2020.1734922.

Lane, Jason E., ed. *Building a Smarter University: Big Data, Innovation, and Analytics*. Albany: State University of New York Press, 2014.

Lane, Jason, and B. Alex Finsel. "Fostering Smarter Colleges and Universities: Data, Big Data, and Analytics." In *Building a Smarter University: Big Data, Innovation, and Analytics*, edited by Jason Lane, 3–26. Albany: State University of New York Press, 2014.

Langlois, Ganaele, and Greg Elmer. "Impersonal Subjectivation from Platforms to Infrastructures." *Media, Culture & Society* 41, no. 2 (2018): 236–251. https://doi.org/10.1177/0163443718818374.

Lascoumes, Pierre, and Patrick Le Galès. "Understanding Public Policy through Its Instruments—from the Nature of Instruments to the Sociology of Public Policy Instrumentation." *Governance: An International Journal of Policy, Administration, and Institutions* 20, no. 1 (2007): 1–21.

Lee, Philip. "The Curious Life of in Loco Parentis in American Universities." *Higher Education in Review* 8 (2011): 65–90. https://scholar.harvard.edu/files/philip_lee/files/vol8lee.pdf.

Lee, Robert, and Tristan Ahtone. "Land-Grab Universities." *High Country News*, March 30, 2020. https://www.hcn.org/issues/52.4/indigenous-affairs-education-land-grab-universities.

Lee, Robert, Tristan Ahtone, Margaret Pearce, Kalen Goodluck, Geoff McGhee, Cody Leff, Katherine Lanpher, and Taryn Salinas. "Land Grab Universities." *High Country News*. Accessed October 19, 2023. https://www.landgrabu.org/.

Leigh, Johanna, Sarah Dunnett, and Lisa Jackson. "Predictive Police Patrolling to Target Hotspots and Cover Response Demand." *Annals of Operations Research* 283 (2019): 395–410. https://doi.org/10.1007/s10479-017-2528-x.

Levine, Alexandra S. "Suicide Hotline Shares Data with For-Profit Spinoff, Raising Ethical Questions." *Politico*, January 28, 2022. https://www.politico.com/news/2022/01/28/suicide-hotline-silicon-valley-privacy-debates-00002617.

Lichtenstein, Nelson. *The Retail Revolution: How Wal-Mart Created a Brave New World of Business*. New York: Metropolitan, 2009.

Lindabary, Jasmine R., and Danielle J. Corple. "Privacy for Whom? A Feminist Intervention in Online Research Practice." *Information, Communication & Society* 22, no. 10 (2019): 1458–1459. https://doi.org/10.1080/1369118X.2018.1438492.

Lippert, Katherine. "Amid Coronavirus, USC Is Requiring Facial Recognition Scans of Students Living on Campus, but the Technology Sparks Controversy." *USC Annenberg Media*, May 15, 2020. https://www.uscannenbergmedia.com/2020/05/15/amid-coronavirus-usc-is-requiring-facial-recognition-scans-of-students-living-on-campus-but-the-technology-sparks-controversy/.

Lipson, Sarah Ketchen, Adam Kern, Daniel Eisenberg, and Alfiee M. Breland-Noble. "Mental Health Disparities among College Students of Color." *Journal of Adolescent Health* 63, no. 3 (2018): 348–56. https://doi.org/10.1016/j.jadohealth.2018.04.014

Lipson, Sarah Ketchen, Emily G. Lattie, and Daniel Eisenberg. "Increased Rates of Mental Health Service Utilization by U.S. College Students: 10-Year Population-Level Trends (2007–2017)." *Psychiatric Services* 70, no. 1 (2018): 60–63. https://doi.org/10.1176/appi.ps.201800332.

Lipson, Sarah Ketchen, Julia Raifman, Sara Abelson, and Sari L. Reisner. "Gender Minority Mental Health in the U.S.: Results of a National Survey on College Campuses." *American Journal of Preventative Medicine* 57, no. 3 (2019): 293–301. https://doi.org/10.1016/j.amepre.2019.04.025.

Liu, Cindy H., Courtney Stevens, Sylvia H. M. Wong, Miwa Yasui, and Justin A. Chen. "The Prevalence and Predictors of Mental Health Diagnoses and Suicide among U.S. College Students: Implications for Addressing Disparities in Service Use." *Depression & Anxiety* 36, no. 1 (2019): 8–17. https://doi.org/10.1002/da.22830.

Logan, Charles. "Towards Abolishing Online Proctoring: Counter-Narratives, Deep Change, and Pedagogies of Educational Dignity." *Journal of Interactive Technology & Pedagogy*, December 2021. https://jitp.commons.gc.cuny.edu/toward-abolishing-online-proctoring-counter-narratives-deep-change-and-pedagogies-of-educational-dignity/.

Lorde, Audre. *A Burst of Light and Other Essays*. London: Sheba Feminist, 1988.

Lubar, Steven. "'Do Not Fold, Spindle or Mutilate': A Cultural History of the Punch Card." *Journal of American Culture* 15, no. 4 (1992): 43–55.

Luitse, Dieuwertje, and Wiebke Denkena. "The Great Transformer: Examining the Role of Large Language Models in the Political Economy of AI." *Big Data & Society* 8, no. 2 (July 2021): 1–14. https://doi.org/10.1177/20539517211047734.

Lum, Kristian, and William Isaac. "To Predict and Serve?" *Significance* 13, no. 5 (2016): 14–19. https://doi.org/10.1111/j.1740-9713.2016.00960.x.

Lupton, Deborah. *Digital Sociology*. London: Routledge, 2015.

Lupton, Deborah. "The Diverse Domains of Quantified Selves: Self-Tracking Modes and Dataveillance." *Economy and Society* 45, no. 1 (2016): 101–122. https://doi.org/10.1080/03085147.2016.1143726.

Lupton, Deborah. *The Quantified Self*. Cambridge: Polity Press, 2016.

Lyon, David. *The Culture of Surveillance: Watching as a Way of Life*. New York: John Wiley & Sons, 2018.

Maass, Dave. "Scholars under Surveillance: How Campus Police Use High Tech to Spy on Students." Electronic Frontier Foundation, March 9, 2021. https://www.eff.org/deeplinks/2021/03/scholars-under-surveillance-how-campus-police-use-high-tech-spy-students#social.

Maass, Dave, and Mike Katz-Lacabe. "Alameda and Contra Costa County Sheriffs Flew Drones Over Protests." Electronic Frontier Foundation, December 5, 2018. https://www.eff.org/deeplinks/2018/12/alameda-and-contra-costa-county-sheriffs-flew-drones-over-protests.

Madoré, Marianna, Anna Zeemont, Joaly Burgos, Jane Guskin, Hailey Lam, and Andréa Stella. "Resisting Surveillance, Practicing/Imagining the End of Grading," *Journal of Interactive Technology & Pedagogy*, December 10, 2021. https://jitp.commons.gc.cuny.edu/resisting-surveillance-practicing-imagining-the-end-of-grading/.

Mahelona, Keoni, Gianna Leoni, Suzanne Duncan, and Miles Thompson. "OpenAI's Whisper Is Another Case Study in Colonialism." *PapaReo* (blog), January 24, 2023. https://blog.papareo.nz/whisper-is-another-case-study-in-colonisation/.

Mann, Monique, Peta Mitchell, Marcus Foth, and Irina Anastasiu. "#BlockSidewalk to Barcelona: Technological Sovereignty and the Social License to Operate Smart Cities." *Journal of the Association for Information Science and Technology* 71, no. 9 (2020): 995–1141. https://doi.org/10.1002/asi.24387.

Marachi, Roxana, and Lawrence Quill. "The Case of Canvas: Longitudinal Datafication through Learning Management Systems." *Teaching in Higher Education* 25, no. 4 (2020): 418–434. https://doi.org/10.1080/13562517.2020.1739641.

Mattelart, Armand. *The Information Society*. London: Sage, 2003.

McCanne, Michael. "When Polaroid Workers Fought Apartheid." *Dissent*, August 14, 2020. https://www.dissentmagazine.org/online_articles/when-polaroid-workers-fought-apartheid/.

McGee, Ebony O., and David Stovall. "Reimagining Critical Race Theory in Education: Mental Health, Healing, and the Pathway to Liberatory Praxis." *Educational Theory* 65, no. 5 (2015): 491–511. https://doi.org/10.1111/edth.12129.

McGregor, Andrew. "Black Labor, White Profits, and How the NCAA Weaponized the Thirteenth Amendment." *Sport in American History*, March 1, 2018. https://ussporthistory.com/2018/03/01/black-labor-white-profits-and-how-the-ncaa-weaponized-the-thirteenth-amendment/.

McKenzie, Lindsay. "Secure File Sharing Compromises University Security." *Inside Higher Ed*, April 6, 2021. https://www.insidehighered.com/news/2021/04/07/accellion-data-security-breach-latest-hit-universities

Mercury. "'Alexa, Why Are You Here?'" April 29, 2019. https://utdmercury.com/alexa-where-are-you-here/.

Mertz, Emily. "University of Alberta Students Get in Touch with Well-Being, Mental Health through WellTrack App." *Global News*, November 21, 2018. https://globalnews.ca/news/4686918/mental-health-university-of-alberta-students-welltrack-app/.

Miller, Peter, and Nikolas S. Rose. *Governing the Present: Administering Economic, Social and Personal Life*. Cambridge: Polity, 2008.

Min-Allah, Nasro, and Saleh Alrashed. "Smart Campus—a Sketch." *Sustainable Cities and Society* 59 (2020): 1–15. https://doi.org/10.1016/j.scs.2020.102231.

Mittelstadt, Jennifer. "Building a New Framework of Values for the University: Emerging from the Ivory Tower's Shadow; Interview with Davarian L. Baldwin." *Academe*, Fall 2022. https://www.aaup.org/article/building-new-framework-values-university#.ZGkOtnbMKUk/.

Moll, Ian. "The Fourth Industrial Revolution: A New Ideology." *Triple C: Communication, Capitalism & Critique* 20, no. 1 (February 2022): 45–61. https://doi.org/10.31269/triplec.v20i1.1297.

Montag, Ali. "This University Is Putting Amazon Echo Speakers in Every Dorm Room." CNBC, August 21, 2018. https://www.cnbc.com/2018/08/21/this-university-is-putting-amazon-echo-speakers-in-every-dorm-room.html.

Moore, Dawn. "The Benevolent Watch: Therapeutic Surveillance in Drug Court." *Theoretical Criminology* 15, no. 3 (2011): 255–268. https://doi.org/10.1177/1362480610396649.

Moore, Phoebe V. *The Quantified Self in Precarity: Work, Technology and What Counts*. New York: Routledge, 2018.

Morgan, Jamie. "Will We Work in Twenty-First Century Capitalism? A Critique of the Fourth Industrial Revolution Literature." *Economy and Society* 48, no. 3 (2019): 371–398. https://doi.org/10.1080/03085147.2019.1620027.

Morozov, Evgeny. *To Save Everything Click Here: Technology, Solutionism and the Urge to Fix Problems That Don't Exist*. London: Penguin, 2013.

Musser, Amber Jamilla. "Specimen Days: Diversity, Labor, and the University." *Feminist Formations* 27, no. 3 (2015): 1–20. https://www.jstor.org/stable/43860813.

Nadler, Bella. "Students Call for UCLA Administration to Abolish Campus Police Presence." *Fem Magazine*, June 6, 2021. https://femmagazine.com/students-call-for-ucla-administration-to-abolish-campus-police-presence/.

National Student Clearinghouse Research Center. "Completing College—National by Race and Ethnicity—2017." *National Student Clearinghouse Research Center*, https://nscresearchcenter.org/signaturereport12-supplement-2/.

National Student Clearinghouse Research Center. "Current Term Enrollment Estimates: Fall 2022 Expanded Edition." Last modified February 2, 2023. https://nscresearchcenter.org/current-term-enrollment-estimates/.

Niemtus, Sofia. "Are University Campuses Turning into Mini Smart Cities?" *Guardian*, February 22, 2019. https://www.theguardian.com/education/2019/feb/22/are-university-campuses-turning-into-mini-smart-cities.

Nielsen, Kelly, Laura T. Hamilton, and Veronica Lerma. "Policing College Campuses: The Production of Racialized Risk." *UF Law Faculty Publications* (2020): 1–39. https://doi.org/10.31235/osf.io/pe8my.

Newfield, Christopher. *The Great Mistake: How We Wrecked Public Universities and How We Can Fix Them*. Baltimore: John Hopkins University Press, 2016.

Newfield, Christopher. "What Is New about the New American University?" *Los Angeles Review of Books*, April 5, 2015. https://lareviewofbooks.org/article/new-new-american-university/.

Ochigame, Rodrigo. "The Invention of 'Ethical AI': How Big Tech Manipulates Academia to Avoid Regulation." *Intercept*, December 20, 2019. https://theintercept.com/2019/12/20/mit-ethical-ai-artificial-intelligence/?comments=1.

Ochigame, Rodrigo. "The Long History of Algorithmic Fairness." *Phenomenal World*, January 30, 2020. https://www.phenomenalworld.org/analysis/long-history-algorithmic-fairness/.

Okechukwu, Amaka. *To Fulfill These Rights: Political Struggle over Affirmative Action & Open Admissions*. New York: Columbia University Press, 2019.

Oliver, Indigo. "Inside Lockheed Martin's Sweeping Recruitment on College Campuses." *In These Times*, August 11, 2022. https://inthesetimes.com/article/lockheed-martin-recruiting-military-industrial-complex-student-debt.

Our Data Bodies. "Who We Are." Accessed May 21, 2023. https://www.odbproject.org/about-us-2/.

Ovetz, Robert. "The Algorithmic University: On-line Education, Learning Management Systems, and the Struggle over Academic Labor." *Critical Sociology* 47, nos. 7/8 (2021): 1065–1084. https://doi.org/10.1177/0896920520948931.

Pangrazio, Luciana. "Reconceptualising Critical Digital Literacy." *Discourse: Studies in the Cultural Politics of Education* 37, no. 2 (2016): 163–174. http://dx.doi.org/10.1080/01596306.2014.942836.

Pardo-Guerra, Juan Pablo. *The Quantified Scholar: How Research Evaluations Transformed the British Social Sciences*. New York: Columbia University Press, 2022.

Park, Michael K. "'Stick to Sports'? First Amendment Values and Limitations to Student-Athlete Expression." *Journalism & Mass Communication Quarterly* 99, no. 2 (June 2022): 515–537. https://doi.org/10.1177/10776990211018757.

Parvin, Nassim, and Anne Pollack. "Unintended by Design: On the Political Uses of 'Unintended Consequences.'" *Engaging Science, Technology, and Society* 6 (2020): 320–327. https://doi.org/10.17351/ests2020.497.

Pasquale, Frank. *The Black Box Society: The Secret Algorithms That Control Money and Information*. Cambridge, MA: Harvard University Press, 2015.

Patel, Vimal. "Are Students Socially Connected? Check Their Dining-Hall-Swipe Data." *Chronicle of Higher Education*, April 9, 2019. https://www.chronicle.com/article/are-students-socially-connected-check-their-dining-hall-swipe-data/.

Paul, Kari. "'Ban This Technology': Students Protest US Universities' Use of Facial Recognition." *Guardian*, March 2, 2020. https://www.theguardian.com/us-news/2020/mar/02/facial-recognition-us-colleges-ucla-ban.

PBS Frontline. "Where Did the Test Come From? Americans Instrumental in Establishing Standardized Tests." Accessed October 19, 2023. https://www.pbs.org/wgbh/pages/frontline/shows/sats/where/three.html.

People's Dispatch. "US Supreme Court Strikes Down Student Debt Relief—Another Blow to Education for All." June 30, 2023. https://peoplesdispatch.org/2023/06/30/us-supreme-court-strikes-down-student-debt-relief-another-blow-to-education-for-all/.

Perrigo, Billy. "Exclusive: OpenAI Used Kenyan Workers on Less Than $2 Per Hour to Make ChatGPT Less Toxic." *Time*, January 18, 2023. https://time.com/6247678/openai-chatgpt-kenya-workers/.

Phan, Thanh Long. "How ChatGPT Is Fine-Tuned Using Reinforcement Learning." *Dida*, April 11, 2023. https://dida.do/blog/chatgpt-reinforcement-learning.

Powles, Julia, and Helen Nissenbaum. "The Seductive Diversion of 'Solving Bias' in Artificial Intelligence." *OneZero*, December 7, 2019, https://onezero.medium.com/the-seductive-diversion-of-solvingbias-in-artificial-intelligence-890df5e5ef53.

Predi Analytics. "Services." Last accessed October 1, 2023, https://www.predianalytics.com/services.

Prensky, Marc. "Digital Natives, Digital Immigrants." *On the Horizon* 9, no. 5 (2001): 1–6. http://dx.doi.org/10.1108/10748120110424816.

Proctorio. "Online Proctoring." Accessed May 19, 2021. https://proctorio.com/products/online-proctoring.

Quinn, Annalisa. "Book News: Amazon Wants to Ship Products before You Even Buy Them." *The Two-Way: Breaking News from NPR*, January 20, 2014. http://www.npr.org/sections/thetwo-way/2014/01/20/264187990/book-news-amazon-wants-to-ship-products-before-you-even-buy-them.

Raley, Rita. "Dataveillance and Counterveillance." In *Raw Data Is an Oxymoron*, edited by Lisa Gitelman, 121–145. Cambridge, MA: MIT Press, 2013.

Randazzo, Marisa R., and J. Kevin Cameron. "From Presidential Protection to Campus Security: A Brief History of Threat Assessment in North American Schools and Colleges." *Journal of College Student Psychotherapy* 26, no. 4 (2012): 277–290. https://doi.org/10.1080/87568225.2012.711146.

Rhue, Lauren. "Racial Influence on Automated Perceptions of Emotions." *Social Science Research Network*, November 2018. https://papers.ssrn.com/sol3/papers.cfm?abstract_id=3281765.

Richardson, Rashida, Jason Schultz, and Kate Crawford. "Dirty Data, Bad Predictions: How Civil Rights Violations Impact Police Data, Predictive Policing Systems, and Justice." *New York University Law Review Online* 94, no. 192 (March 5, 2019). https://papers.ssrn.com/sol3/papers.cfm?abstract_id=3333423#.

Ricker Schulte, Stephanie. "Personalization." In *Digital Keywords*, 242–255. Princeton, NJ: Princeton University Press, 2016.

Roberts, Dorothy E. "Book Review: Digitizing the Carceral State." *Harvard Law Review* 132, no. 6 (2019): 1695–1728.

Robertson, Craig. *The Passport in America: The History of a Document*. Oxford: Oxford University Press, 2010.

Rodríguez, Dylan. "Beyond 'Police Brutality': Racist State Violence and the University of California." *American Quarterly* 64, no. 2 (2012): 301–313. https://doi.org/10.1353/aq.2012.0012.

Roza, Matthew. "Student Fear for Their Data Privacy after University of California Invests in Private Equity Firm." *Salon*, July 28, 2020. https://

www.salon.com/2020/07/28/students-fear-for-their-data-privacy-after-university-of-california-invests-in-private-equity-firm/.

Rubel, Alan, and Kyle M. L. Jones. "Student Privacy in Learning Analytics: An Information Ethics Perspective." *Information Society* 32, no. 2, 143–159. https://doi.org/10.1080/01972243.2016.1130502.

Rubley, Julie. "Achieving the Smart Campus of Tomorrow Today." *Educause Review*, November 25, 2019. https://er.educause.edu/blogs/sponsored/2019/11/achieving-the-smart-campus-of-tomorrow-today.

Rosner, Rachel I. "Manualizing Psychotherapy: Aaron T. Beck and the Origins of Cognitive Therapy of Depression." *European Journal of Psychotherapy & Counselling* 20, no. 1 (2018): 25–47. https://doi.org/10.1080/13642537.2017.1421984.

Sacks, Ethan. "Alexa Privacy Fail Highlights Risks of Smart Speakers." NBC News, May 26, 2018. https://www.nbcnews.com/tech/innovation/alexa-privacy-fail-highlights-risks-smart-speakers-n877671.

Sadowski, Jathan. *Too Smart: How Digital Capitalism Is Extracting Data, Controlling Our Lives, and Taking Over the World*. Cambridge, MA: MIT Press, 2020.

Saha, Somen. "World's First Voice-Enabled Student Experience by N-Powered .com." YouTube video. Posted by Somen Saha, January 30, 2018. https://youtu.be/wtdroQzWRhY.

Saint Louis University. "Welcome to SLU, Alexa." YouTube video. Posted by SaintLouisUniversity, August 9, 2018. https://youtu.be/Q8jfcxMYbQw.

Sanderson, Jimmy, Blair Browning, and Annelie Schmittel. "Education on the Digital Terrain: A Case Study Exploring College Athletes' Perceptions of Social-Media Training." *International Journal of Sport Communication* 8, no. 1 (2015):103–124. https://doi.org/10.1123/IJSC.2014-0063.

Saravanan, Raja. "Facial Recognition Can Give Students Better Service (and Security)." *Ellucian* (blog). Accessed May 19, 2023. https://www.ellucian.com/blog/facial-recognition-campus-benefits-security-risks.

Savio, Mario. "Sit-In Address on the Steps of Sproul Hall." Speech delivered at the University of California, Berkeley, December 2, 1964.

Schiff, Daniel. "Education for AI, *not* AI for Education: The Role of Education and Ethics in National AI Policy Strategies." *International Journal of Artificial Intelligence in Education* 32 (2022): 527–563. https://doi.org/10.1007/s40593-021-00270-2.

Schiller, Dan. *Digital Depression: Information Technology and Economic Crisis*. Champaign: University of Illinois Press, 2014.

Schneider, Jack, and Ethan Hutt, "Making the Grade: A History of the A–F Marking Scheme." *Journal of Curriculum Studies* 46, no. 2 (2014): 201–224. http://dx.doi.org/10.1080/00220272.2013.790480.

Schwarz, John. "The Origin of Student Debt: The Danger of Educated Proles." *Intercept*, August 25, 2022. https://theintercept.com/2022/08/25/student-loans-debt-reagan/.

Scott-Clayton, Judith, and Jing Li. "Black-White Disparity in Student Loan Debt More Than Triples after Graduation." *Brookings*, October 20, 2016. https://www.brookings.edu/research/black-white-disparity-in-student-loan-debt-more-than-triples-after-graduation/.

Seamster, Louise, and Raphaël Charron Chénier. "Predatory Inclusion and Education Debt: Rethinking the Racial Wealth Gap." *Social Currents* 4, no. 3 (2017): 199–207. https://doi.org/10.1177/2329496516686620.

Selwyn, Neil. *Digital Technology and the Contemporary University: Degrees of Digitization*. London: Routledge, 2014.

Selwyn, Neil. *Education and Technology: Key Issues and Debates*. 2nd ed. New York: Bloomsbury, 2016.

Shenkoya, Temitayo, and Euiseok Kim. "Sustainability in Higher Education: Digital Transformation of the Fourth Industrial Revolution and Its Impact on Open Knowledge." *Sustainability* 15, no. 3 (2023): 1–16. https://doi.org/10.3390/su15032473.

Shrader, Stuart. *Badges without Borders: How Global Counterinsurgency Transformed American Policing*. Berkeley: University of California Press, 2019.

Siemens. "Attractive Higher Education Campuses." Last modified August 2022. https://assets.new.siemens.com/siemens/assets/api/uuid:fb4dad9c-5c51-43aa-96ce-8d64128d1a44/smart-campus-university-ipdf-en.pdf.

Silbey, Susan S. "How Not to Teach Ethics." *MIT Faculty Newsletter*, September/October 2018. http://web.mit.edu/fnl/volume/311/silbey.html.

Simon, Caroline. "More and More Students Need Mental Health Services: But Colleges Struggle to Keep Up." *USA Today*, May 4, 2017. https://www.usatoday.com/story/college/2017/05/04/more-and-more-students-need-mental-health-services-but-colleges-struggle-to-keep-up/37431099/.

Singer, Natasha. "They Loved Your G.P.A. Then They Saw Your Tweets." *New York Times*, November 9, 2013. https://www.nytimes.com/2013/11/10/business/they-loved-your-gpa-then-they-saw-your-tweets.html.

Slaughter, Sheila, and Gary Rhoades. "The Neo-liberal University." *New Labor Forum*, no. 6 (2000): 73–79. https://www.jstor.org/stable/40342886.

Sloan, John J., III. "The Modern Campus Police: An Analysis of their Evolution, Structure, and Function." *American Journal of Police* 11, no. 1 (1992): 85–104.

Sloan, John Jay, III. "Race, Violence, Justice, and Campus Police." *Footnotes: A Publication of the American Sociological Association* 48, no. 4 (2020): 9–12. https://www.asanet.org/news-events/footnotes/jul-aug-2020/features/race-violence-justice-and-campus-police.

Sloan, John Jay, III, and Bonnie S. Fisher. *The Dark Side of the Ivory Tower: Campus Crime as a Social Problem*. Cambridge: Cambridge University Press, 2010.

Smith, Erika E. "The Digital Native Debate in Higher Education: A Comparative Analysis of Recent Literature." *Canadian Journal of Learning and Technology* 38, no. 3 (2012): 1–18.

Smith, Michael Clay. "The Ancestry of Campus Violence." *New Directions for Student Services* 1989, no. 4 (1989): 5–15. https://doi.org/10.1002/ss.37119894703.

So, Mirai, Sosei Yamaguchi, Sora Hashimoto, Mitsuhiro Sado, Furukawa A. Toshi, and Paul McCrone. "Is Computerised CBT Really Helpful for Adult Depression? A Meta-analytic Re-evaluation of CCBT for Adult Depression in Terms of Clinical Implementation and Methodological Validity." *BMC Psychiatry*, no. 13 (2013): 113–127. https://doi.org/10.1186/1471-244X-13-113.

Social Sentinel. "Our Solution." Accessed May 18, 2020. https://navigate360.com/solutions/social-sentinel-scanning-businesses/.

Stanley, J. Adrian. "UCCS Secretly Photographed Students to Advance Facial Recognition Technology." *Colorado Springs Independent*. Last modified October 15, 2020. https://www.csindy.com/temporary_news/uccs-secretly-photographed-students-to-advance-facial-recognition-technology/article_f314793e-20aa-586c-a0e9-db2cadd394a4.html.

Stanley, Jay. "Four Problems with the ShotSpotter Gunshot Detection System." ACLU. Last modified October 14, 2021. https://www.aclu.org/news/privacy-technology/four-problems-with-the-shotspotter-gunshot-detection-system.

Stevenson, Megan, and Sandra G. Mayson. "The Scale of Misdemeanor Justice." *Boston University Law Review* 98, no. 731 (2018): 769–770. https://scholarship.law.upenn.edu/faculty_scholarship/2391.

Stone, Karen J., and Edward N. Stoner II. "Revisiting the Purpose and Effect of FERPA." Stetson University College of Law 23rd Annual National Conference on Law and Higher Education (February 2002). http://www.stetson.edu/law/academics/highered/home/media/2002/Revisiting_the_Purpose_of_FERPA.pdf.

Students for Fair Admissions v. President and Fellows of Harvard College, No. 20–1199, slip op. (U.S. Oct. 31, 2022), https://www.supremecourt.gov/opinions/22pdf/20-1199_hgdj.pdf.

Sunstein, Cass. "Nudging: A Very Short Guide." *Journal of Consumer Policy* 37 (2014): 583–588. https://doi.org/10.1007/s10603-014-9273-1.

Supiano, Beckie. "Nudging Looked Like It Could Help Solve Key Problems in Higher Ed: Now That's Not So Clear." *Chronicle of Higher Education*,

September 4, 2019. https://www.chronicle.com/article/nudging-looked-like-it-could-help-solve-key-problems-in-higher-ed-now-thats-not-so-clear/.

Sutherland, Chanel M. "Digital Natives: 4 Ways Technology Has Changed 'the' Student." *Explorance*, March 31, 2016. https://explorance.com/blog/digital-natives-4-ways-technology-changed-student/.

Suvin, Mary Clotilda. "Everything You Need to Know about an Effective Faculty Management Solution." Creatrix Campus. Last modified December 12, 2022. https://www.creatrixcampus.com/blog/everything-you-need-know-about-effective-faculty-management-solution.

Swauger, Shea. "Our Bodies Encoded: Algorithmic Test Proctoring in Higher Education." *Hybrid Pedagogy*, April 2, 2020. https://hybridpedagogy.org/our-bodies-encoded-algorithmic-test-proctoring-in-higher-education/.

Swauger, Shea. "Software That Monitors Students during Tests Perpetuates Inequality and Violates Their Privacy." *MIT Technology Review*, August 7, 2020. https://www.technologyreview.com/2020/08/07/1006132/software-algorithms-proctoring-online-tests-ai-ethics/.

Sweeney, Latanya. "Discrimination in Online Ad Delivery." *Communications of the ACM* 56, no. 5 (2013): 44–54. https://doi.org/10.1145/2447976.2447990.

Taiz, Lincoln, Joel Yellin, Deanna Shemek, Karen Bassi, and Tony Tromba. "UCSC Admission: History and Opportunities." Accessed May 19, 2021. https://senate.ucsc.edu/archives/Past%20Issues/narrative-evaluations/AdmissionsHisandOpp_doc.pdf.

Taylor, Astra, and Jathan Sadowski. "How Companies Turn Your Facebook Activity into a Credit Score." *Nation*, May 27, 2015. https://www.thenation.com/article/archive/how-companies-turn-your-facebook-activity-credit-score/.

Teräs, Marko, Juha Suoranta, Hanna Teräs, and Mark Curcher. "Post-Covid-19 Education and Education Technology 'Solutionism': A Seller's Market." *Postdigital Science and Education* 2 (July 2020): 863–878. https://doi.org/10.1007/s42438-020-00164-x.

Thornhill, Ted. "We Want Black Students, Just Not You: How White Admissions Counselors Screen Black Prospective Students." *Sociology of Race and Ethnicity* 5, no. 4 (October 2019): 456–470. https://doi.org/10.1177/2332649218792579.

Torbin, Ariana. "HUD Sues Facebook over Housing Discrimination and Says the Company's Algorithms Have Made the Problem Worse." *ProPublica*. Last modified March 18, 2019. https://www.propublica.org/article/hud-sues-facebook-housing-discrimination-advertising-algorithms.

Tulumello, Simone, and Fabio Iapaolo. "Policing the Future, Disrupting Urban Policy Today: Predictive Policing, Smart City, and Urban Policy in

Memphis (TN)." *Urban Geography* 43, no. 3 (2021): 448–469. https://doi.org/10.1080/02723638.2021.1887634.

Uskov, Vladimir L., Jeffrey P. Bakken, Robert James Howlett, and Lakhmi C. Jain, eds. *Smart Universities: Concepts, Systems, and Technologies.* New York: Springer International, 2018.

Vallance, Chris. "Artificial Intelligence Could Lead to Extinction, Experts Warn." BBC News, May 30, 2023. https://www.bbc.com/news/uk-65746524.

Vanderslot, Jodie. "Mental Health through WellTrack." *Excalibur.* Last modified December 1, 2017. https://www.excal.on.ca/health/2017/12/01/mental-health-through-welltrack/.

Varian, Hal. "Differential Pricing and Efficiency." *First Monday* 1, no. 2 (1996). http://firstmonday.org/ojs/index.php/fm/article/view/473/394.

Viberg, Olga, Mathias Hatakka, Olof Balter, and Anna Mavroudi. "The Current Landscape of Learning Analytics in Higher Education." *Computers in Human Behavior* 89 (December 2018): 98–110. https://doi.org/10.1016/j.chb.2018.07.027.

Walsh, James P. "Education or Enforcement? Enrolling Universities in the Surveillance and Policing of Migration." *Crime, Law, and Social Change* 71 (2019): 325–344. https://doi.org/10.1007/s10611-018-9792-9.

Wang, Marian. "Public Universities Ramp Up Aid for the Wealthy, Leaving the Poor Behind." *ProPublica,* September 11, 2013. https://www.propublica.org/article/how-state-schools-ramp-up-aid-for-the-wealthy-leaving-the-poor-behind/.

Watkins, Grace. "'Cops Are Cops': American Campus Police and the Global Carceral Apparatus." *Comparative American Studies* 19, nos. 3/4 (2020): 242–256. https://doi.org/10.1080/14775700.2021.1895039.

Watters, Audrey. "Ed-Tech and the Californian Ideology." *Hack Education,* May 17, 2015. http://hackeducation.com/2015/05/17/ed-tech-ideology.

Watters, Audrey. *Teaching Machines: The History of Personalized Learning.* Cambridge, MA: MIT Press, 2021.

Weagle, Stephanie. "From Safe Campus to Smart Campus." *Campus Security Today,* April 1, 2019. https://campuslifesecurity.com/Articles/2019/04/01/From-Safe-Campus-to-Smart-Campus.aspx?Page=1.

Weinberg, Lindsay. "Rethinking Fairness: An Interdisciplinary Survey of Critiques of Hegemonic Fairness Approaches." *Journal of Artificial Intelligence Research* 74 (2022): 75–109. https://doi.org/10.1613/jair.1.13196.

WellTrack. "Clinicians & Administrators." Accessed January 30, 2019. https://welltrack.com/for-higher-ed/.

West, Sarah Myers, Meredith Whittaker, and Kate Crawford. "Discriminating Systems: Gender, Race, and Power in AI." AI Now (2019). https://ainowinstitute.org/publication/discriminating-systems-gender-race-and-power-in-ai-2.

West Virginia University. "WVU Modernization Program." Accessed October 3, 2023. https://strategicinitiatives.wvu.edu/wvu-modernization-program/about.

White v. Davis. Vol. 13, Cal. 3d, March 24, 1975. https://law.justia.com/cases/california/supreme-court/3d/13/757.html.

White, James Merricks. "Anticipatory Logics of the Smart City's Global Imaginary." *Urban Geography* 37, no. 4 (2016): 572–589. https://doi.org/10.1080/02723638.2016.1139879.

Whitman, Madisson. "'We Called That a Behavior': The Making of Institutional Data." *Big Data & Society* 7, no. 1 (2020): 1–13. https://doi.org/10.1177/2053951720932200.

Whittaker, Meredith. "The Steep Cost of Capture." *Interactions* 18, no. 6 (2021): 50–55. https://doi.org/10.1145/3488666.

Wilder, Craig Steven. *Ebony & Ivy: Race, Slavery, and the Troubled History of America's Universities*. New York: Bloomsbury Press, 2013.

Wilkinson, James. "UC Davis Paid at Least $175,000 to Hide References to Infamous 2011 'Pepper-Spray Incident' from Google Searches." *Daily Mail*, last modified April 14, 2016. https://www.dailymail.co.uk/news/article-3540299/UC-Davis-chancellor-Lind-Katehi-fire-taxpayer-funded-uni-paid-175-000-hide-references-infamous-2011-pepper-spray-incident-Google-searches-university.html.

Williams, Corbretti D. "Race and Policing in Higher Education." *Activist History Review*, November 19, 2019. https://activisthistory.com/2019/11/19/race-and-policing-in-higher-education/.

Williams, Frankwood E. "Mental Hygiene and the College Student," *Mental Hygiene* 5 (1921): 283–301.

Williams, Krystal L., and BreAnna L. Davis. "Public and Private Investments and Divestments in Historically Black Colleges and Universities." *American Council of Education Issue Brief* (2019): 1–10. https://vtechworks.lib.vt.edu/bitstream/handle/10919/89184/PublicPrivateHbcus.pdf.

Williamson, Ben. "Digital Education Governance: Data Visualization, Predictive Analytics, and 'Real-Time' Policy Instruments." *Journal of Education Policy* 31, no. 2 (April 2015): 123–141. https://doi.org/10.1080/02680939.2015.1035758.

Williamson, Ben. "Psychodata: Disassembling the Psychological, Economic, and Statistical Infrastructure of 'Socio-emotional Learning.'" *Journal of Education Policy* 36, no. 1 (2021): 129–154. https://doi.org/10.1080/02680939.2019.1672895.

Williamson, Ben. "The Hidden Architecture of Higher Education: Building a Big Data Infrastructure for the 'Smarter University.'" *International Journal of Higher Education* 15, no. 12 (2018): 1–26. https://doi.org/10.1186/s41239-018-0094-1.

Williamson, Kimberly, and Rene F. Kizilcec. "A Review of Learning Analytics Dashboard Research in Higher Education: Implications for Justice, Equity, Diversity, and Inclusion." *LAK22: 12th International Learning Analytics and Knowledge Conference (Association for Computing Machinery)* (2022): 260–270. https://doi.org/10.1145/3506860.3506900.

Winfield, Ann Gibson. *Eugenics and Education in America: Institutionalized Racism and the Implications of History, Ideology, and Memory.* New York: Peter Lang, 2007.

Winner, Langdon. "Do Artifacts Have Politics." *Daedalus* 109, no. 1 (Winter 1980): 121–136.

Wolters, Raymond. *The New Negro on Campus: Black College Rebellions of the 1920s.* Princeton, NJ: Princeton University Press, 1975.

Wriggins, Jennifer. "The Color of Property and Auto Insurance: Time for Change." *Florida State University Law Review* 49, no. 203 (2022): 203–256.

Wu, Xiaolin, and Xi Zhang. "Automated Inference on Criminality Using Face Images." *arXiv* (2016): 1–9. https://arxiv.org/pdf/1611.04135v1.pdf.

Xiang, Chloe. "ChatGPT Can Replace the Underpaid Workers Who Train AI, Researchers Say." *Vice*, March 29, 2023. https://www.vice.com/en/article/ak3dwk/chatgpt-can-replace-the-underpaid-workers-who-train-ai-researchers-say/.

Yee, Georgia. "Protect Student Privacy: A Renewed Call to Action against Proctorio." Last modified October 1, 2020. https://docs.google.com/document/d/1117835S2RQkQN_-Ij8nZ2qEzCKgNWqle-O7onKdzpaA/edit.

Young, Harrison. "UGA Dining Halls to Introduce Iris Scanners." *Red & Black*, last modified April 22, 2017. https://www.redandblack.com/uganews/uga-dining-halls-to-introduce-eye-scanners/article_0b1703e2-2562-11e7-9dac-f38fb6755209.html.

Zalcmanis-Lai, Ophelie. "Exclusive: University Set to Introduce Online Mental Health Platform to Students." *Ryersonian*, April 9, 2016. https://ryersonian.ca/exclusive-university-set-to-introduce-online-mental-health-platform-to-students/.

Zegart, Amy, and Michael Morell. "Spies, Lies, and Algorithms: Why U.S. Intelligence Agencies Must Adapt or Fail." *Foreign Affairs*, April 16, 2019. https://www.foreignaffairs.com/articles/2019-04-16/spies-lies-and-algorithms.

Zeide, Elana. "The Limits of Education Purpose Limitations." *University of Miami Law Review* 71, no. 2 (2017): 494–527. https://repository.law.miami.edu/umlr/vol71/iss2/8.

Zeide, Elana. "Student Privacy Principles for the Age of Big Data: Moving Beyond FERPA and FIPPS." *Drexel Law Review* 8, no. 339 (2016): 339–394.

Zimpher, Nancy L. "Building a Smarter University: Big Data, Innovation, and Ingenuity." In *Building a Smarter University: Big Data, Innovation, and Analytics*, ed. Jason E. Lane, xi–xv. Albany: State University of New York Press, 2014.

Zraick, Karen, and Mihir Zaveri. "Harvard Student Says He Was Barred from U.S. over His Friends' Social Media Posts." *New York Times*, August 27, 2019. https://www.nytimes.com/2019/08/27/us/harvard-student-ismail-ajjawi.html.

Index

Abd-Almageed, Wael, 123
Abdurahman, J. Khadijah, 155
ableism, 21, 79, 117
academic courses, access and selection, 3, 18, 33, 78, 115; massive open online (MOOCs), 4
academic freedom, 45, 67, 74, 77, 140, 162
academic majors, 66, 72, 78
Acoli, Sundiata, 165
admissions, 49–64, 116; fraud and deception in, 48–49; open, 50, 52–53; racial/socioeconomic bias in, 49–50, 51–52, 53–54, 56–57, 61, 64, 116; smart technology approach, 49, 55–57, 61–62
Adorno, Theodor, 29
advertising, targeted, 17, 25, 27, 29–30, 31, 76, 106
advising and advisors, 3, 32–33, 69, 72, 74–75, 77; e-advisors, 18, 70, 78
affirmative action, 50, 51–52, 53, 61, 97, 116
#Against Surveillance teach-in, 158
Ahmed, Sara, 99
AI Now Institute, 148
alerting systems, 38–39, 119
Alexa, 103–7
algorithms, 27, 54–55, 59, 119; in admissions and recruitment, 55, 61; decision-making based on, 116–17; as discrimination tool, 31; in exam proctoring technology, 116–17, 118; in FRT, 119; personalization, 27, 28; in predictive policing, 39, 145; risk assessment, 69, 144; in smart city initiatives, 41; in student retention, 69

Amazon, 3, 12, 42, 59, 105–6, 140, 165; Alexa-enabled devices, 103–7, 106; Echo Dots, 18–19, 89, 103–7; e-commerce and personalization practices, 28, 32, 33, 35; Mechanical Turk, 155; Prime, 105; Web Services, 69–70, 165
American Civil Liberties Union, 122
American College Testing (ACT), 48, 54
American Student Health Association, 86–88
Amherst College, 85–86
Andrejevic, Mark, 28–29
annotation: tools, 158; workers, 155, 156–57
anonymized data, 41–42, 55, 78, 100–102
anthropomorphism, 134, 154
anti-war activism, 10, 131, 164
Aoun, Joseph, 44–45
Arasaratnam, Gaya, 91
Arizona State University, 45, 71, 104, 106, 125
artificial intelligence (AI), 1, 14, 44–45, 153–54, 160; critical digital literacy approach, 152–57; ethics, 117–18; language engines, 129; military/national security applications, 139–40, 165; university research in, 19, 137–49. *See also* emotion recognition technology; facial recognition technology; machine learning; predictive policing
Asian American students, 51

AspirEDu, Instructor Insight, 162–63
assessment, in education: inequalities-based, 114, 117. *See also* exam proctoring software; grading system
audio and voice data, 105–6, 111–12, 153–54
austerity politics, 1, 3–4, 10, 11, 40, 78, 85, 98–99, 110, 159, 160–61, 168–69; austerity consensus, 15–16; data sharing and, 9; grading systems and, 113, 116, 118; impact on faculty, 15; inequities associated with, 13, 113, 152, 167; recruitment and, 35, 49, 57, 58–59; securitization/policing and, 136, 145–46; social movement against, 125, 132, 136; student and faculty opposition to, 20–21, 160, 162, 167; student mental health / wellness and, 85, 95–99, 110
Austin Peay State University, Degree Compass system, 33
automation, opposition to, 5–6, 34
autonomy, 42, 67, 77, 78, 160

Baker, Dominique, 51–52
Baldwin, Davarian, 125, 128–29, 168
Ball State University, 83–84
beacon technology, 22, 70–71
Beck, Aaron T., 92
Beer, David, 89
Beers, Clifford, *A Mind That Found Itself*, 88–89
behavioral data tracking, 1, 3, 13, 17, 18, 39, 67, 84, 110, 159; behavioral control applications, 1, 3, 7, 22–23, 59, 69, 71, 79, 81, 105, 125, 158; of consumers, 25–27; of faculty, 159, 163–64; predictive analytics based on, 17; racial bias, 128–29; in recruitment and admissions, 17, 55, 56, 58, 61. *See also* emotion recognition / tracking technology; exam proctoring software; facial recognition technology
behavioral economics, 24
Benanav, Aaron, 46
Benjamin, Ruha, 96, 147
Beswick, Kim, 35
Biden administration, 167

Big Tech companies, 8, 59, 140, 157, 165
biometric technology, 62, 112, 113, 123–24, 163. *See also* facial recognition technology
Black athletes, 65–67, 121
Blackboard, 74
Black Lives Matter movement, 127
Black Panthers, 165
Black Power movement, 50
Black students, 13, 54, 64, 65–67, 97, 98, 167. *See also* racism / racial inequities
Blue CRUSH, 13–14, 146, 147
Boggs, Abigail, 15, 63, 122, 137, 169
border security, 63, 132, 142–43
Brigham, Carl, 53
Brightspace, 69–70, 74–75
Browne, Simone, 29, 96, 141–42
Buckley, James, 7
Byrd, W. Carson, 53

Caines, Autumm, 156–57
CampusLogic, 57–58
Campus Security Act, 126
Canvas LMS, 19, 74, 76
capitalism, 4, 5–6, 26, 29, 32, 46, 94–95, 107–8, 134, 137, 148, 155
Capture, 54–56; AID tool, 58
carceral institutions, 69–70, 115, 135, 145
carceral technologies, 69–70, 137, 144, 145–46, 149; research in, 13–14, 20–21, 39–40, 137, 138–40, 144–49, 166–67. *See also* Correctional Offender Management Profiling for Alternative Sanctions; predictive policing
card-swipe technology, 22, 77, 123
Cavoukian, Ann, 37
CBT. *See* cognitive behavioral therapy
cell phone tower mimicking devices, 39
CenturyLink, 23
Chan, Ngai Keung, 61–62
chat services, AI-assisted, 100; ChatGPT, 152–57
Chénier, Raphaël Charron, 117–18
Cheong, Pauline Hope, 151
CIA, 139–40

City University of New York Academic Commons, 159
Civics of Technology Conference, 158
Civil Rights Act, 51, 54
civil rights movement, 50, 51, 68, 131
class attendance tracking, 65, 66, 69, 70–71, 77, 120, 163
Clery, Jeanne, 126–27
Clotilda, Mary, 163
cloud technology, 28, 69–70, 74, 75, 165
coalition-based movement, against smart university initiatives, 160–69
Coalition for Critical Technology, 145
cognitive behavioral therapy (CBT), 18, 83, 92–95, 103
Cohen, Sol, 88–89
College and University Machine Records Conference, 124
College and University Systems Exchange (CAUSE), 124
College Boards, 53
college education: cost, 11–12, 78, 167; economic benefits, 64
College Entrance Exam Board, 53, 54
colonialism/colonization, 97, 125, 135, 137, 138, 153–54, 155
Columbia State University, 120
Columbia University, 34
Comeaux, Eddie, 66
community colleges, 11, 57
COMPAS. *See* Correctional Offender Management Profiling for Alternative Sanctions
Computer People for Peace, 165
consent, for data/technology access and use, 7, 9, 99–101, 102, 121, 143, 153–54, 160
Corerain, 71
Corple, Danielle J., 142
corporate-university partnerships, 3, 4, 11, 23, 44–45, 58, 67, 110, 167, 168; in AI research, 19, 138–41, 144; big data socialization and, 107; of HBCUs, 13; of multiversities, 44; for national security, 19; with police, 128, 146; privacy and exploitation issues, 105–8;
for workforce diversity, 13–14; for workforce training, 23, 44–45
corporatization, 14, 45, 140
Correctional Offender Management Profiling for Alternative Sanctions (COMPAS), 59–60, 140, 144–45
Costanza-Chock, Sasha, 84–85
cost of living adjustment (COLA), 133, 136
Cottom, Tressie McMillan, 11
COVID-19 pandemic, 40–41, 54, 70, 75–76, 110, 120, 123, 132–33, 134
Crary, Jonathan, 107
Creatrix Campus, Faculty Management System, 163–64
Crime Awareness and Campus Security Act, 8
criminality, 119. *See also* predictive policing
criminalization, 69, 128, 132, 136, 140–49
criminal legal system, racial and social inequities, 39, 59–60, 141–42; risk assessment algorithms, 69, 144–45. *See also* police/policing; predictive policing
crisis consensus, 15
crisis narratives, 2, 40–41, 126, 145–46
Crisis Text Line, 100
critical digital literacy training, 20, 149, 150–61
critical university studies, 5
Crow, Michael, 45
CyberPsyc Software Solutions, Inc., 83–84, 102, 108, 109
cybersecurity, 13

Dabars, William B., *Designing the New American University*, 45
dashboards, 18, 55, 58, 70, 77, 84
data aggregation, 13, 27, 28, 31–32, 81, 100
data analytics, 8–9, 12, 16–17, 26–27. *See also* algorithms
databases, 27–28, 101
data capture/collection, in higher education, 3, 22–23, 27, 78; corporate policies, 3; as dataveillance, 89, 99–103, 109; history, 6–7; pressures for, 11–12

data doubles, 13, 31
data mining, 8–9, 25, 27
data sharing, 8–9, 100, 129
data storage, 27–28, 55, 75, 78, 123
dataveillance, 26, 99–103
Debt Collective, 167–68
Deloitte, 23, 37–38, 45
democratization, 50, 157
demographic data, 10, 17, 22, 40, 43, 58, 64, 71, 79–80, 146
depersonalization, 6, 36
deregulation, 10–11, 69
desegregation, 52
digital economy, 13–14, 16–17, 20, 25–27, 28–29, 150, 157
digital natives, 19–20, 150–51
digital redlining, 30–31
digital surveillance, 1–2; benefits for students, 65. *See also* biometric technology; emotion recognition technology; facial recognition technology; personalization; securitization
diversity, equity, and inclusion, 58; affirmative action bans and, 116; corporate partnership for, 13–14; harmful forms, 117–18, 142; programs for, 60–61; as recruitment goal, 52–53; of student demographics, 40
Divoky, Diane, 6–7
Donaghue, Ngaire, 102–3
Dong, Zhao Yang, 35
dormitories, 18–19, 89, 103–8, 123
Doty, Philip, 67
drones, 132–33, 139, 165
drug use, 6, 119, 147
Duffy, Brooke Erin, 61–62
Duke University, 141

e-commerce, 16–17, 28, 33–36
economic and social inequalities, 49, 76, 77–78, 113–14, 149; FRT and, 119; grading system and, 114, 116; mental health and, 87–88, 90, 94, 96, 99, 109, 110; personalization and, 29; policing and, 128, 135, 160; precarity, 94–95, 109, 134; recruitment and, 17; retention and, 69; securitization and, 113–14, 149
economic liberalization policies, 10–11
economy, globalized, 52; knowledge economy, 45
Edcom, 124
educational resources, 51–52
educational technology: companies, 45, 75–76, 78, 117, 150, 158; ethics, 158; global market, 75–76
EDUCAUSE, 124
EduNav, 78
Electronic Privacy Information Center, Student Privacy Bill of Rights, 151–52
Ellucian, 119–20
email, 17, 26, 55, 61, 72, 129, 166–67
emotion recognition technology, 71–72, 120–22, 121, 164. *See also* WellTrack self-care tracking app
enrollment, 6, 12, 32–33, 35, 51–52, 55, 56, 80, 130; declines, 11, 40, 51, 54, 78; nontraditional students, 24; post-WW II expansion, 50
ethical issues: AI, 117–18, 140, 157; ChatGPT, 156; data use, 12; educational technologies, 158; information technology, 77–78; personal information collection, 123–24; research, 143; technology ethics education, 164; voice data, 105–6
eugenics movement, 53, 88–89, 114, 119
Evans, Gabe, 134
Examity, 112
exam proctoring software, 20–21, 111–14, 116–17, 118, 158, 166
eye movement monitoring, 111–12

Face++, 121
Facebook, 7, 8, 31
facial recognition technology (FRT), 31, 38, 71–72, 123, 140; in exam proctoring, 111–14, 117–18; in law enforcement, 127, 132–33, 142–43; opposition to, 122; racialized applications, 14, 118–19, 121–22, 141–44

faculty, 85–86, 125, 130, 131, 133, 149, 168; diversity, 127–28, 163–64; exploitation and precarity, 4, 14, 15, 72–73, 82, 159–60, 162–64, 166–67, 168; non-tenure track, 72–73; resistance to smart university initiatives, 20–21, 111, 112, 122, 134, 141, 158, 166–67
Fair Housing Act, 8, 29
Fair Labor Standards Act, 66–67
Family Educational Rights and Privacy Act, 7, 8–9, 104
FBI, 50–51, 132; Joint Terrorism Task Force, 132
feminism / feminist scholarship, 15, 68, 94, 99
Ferguson, Andrew, 148
Ferraro, David, 93
Fiebig, Tobias, 74
Fight for the Future, 122
financial aid, 22, 57–58, 66, 80
financialization, 64, 98
fingerprint identification, 123
First Amendment, 131
first-generation college students, 24, 74
Florida International University, 120
Floyd, George, 149
Ford, Gerald, 7
Ford Foundation, 34–35
for-profit higher education institutions, 10–12, 57, 64, 159
for-profit use, of student data, 8–9, 67, 99, 100–102
Fourth Amendment, 106–7
Frankfurt School, 29
FRT. *See* facial recognition technology

Gaffney, Christopher, 36–37
Galligan, Claire, 122
Galton, Francis, 119
Gandy, Oscar, 26
Gates, Bill, 32
Gates Foundation, 75
Gebru, Timnit, 157
Gegg-Harrison, Whitney, 153
gender discrimination, 29
Georgetown University, 48

Georgia Institute of Technology, 104
Georgia State University, 72, 83–84
Gill, Rosalind, 102–3
Gillespie, Tarleton, 32
Gilliard, Chris, 30, 81
Ginsberg, Olivia, 91
Gitelman, Lisa, 42–43
GitHub, 74
Godrej, Farah, 132
Goff, Jay W., 12–13, 120
Google, 29–30, 157, 165
government-university partnerships, 19, 139, 140–41
GPS technology, 77, 129
grading system, 113, 114–18
graduate students, 19, 83, 133–34, 148–49
graduation, 22–23, 78; rates, 16, 57, 66, 72–73, 108
Graham, Rebecca Dolinsky, 126
grants, 14–15, 18, 57, 75, 124
Great Recession (2007–2009), 50–51, 62, 146
Greenfield, Adam, 37, 43–44
gunshot detection technology, 38, 128–29

Hakanson, David, 103, 107
Haldane, Andrew G., 44
HCL America, 83–84
health information, third party access, 94–95, 101–2
health programs. *See* mental health and wellness self-tracking apps
Henry, Joseph, 126
Hewlett Foundation, 75
Higher Ed Labor United, 167–68
Higher Education Community Vendor Assessment Toolkit, 124
Higher Education Information Security Council, 124
Higher Education Reauthorization Act, 130–31
historically Black colleges and universities (HBCUs), 13, 74, 87, 130, 131
Hong, Sun-ha, 71–72
Horkheimer, Max, 29

housing discrimination, 8, 29, 30–31
Hypothesis (annotation tool), 158

Iapaolo, Fabio, 146–47
IBM, 13–14, 37, 43, 124, 146, 165
identification (ID) cards, 22, 123
identity classification, 79–80
immigrants and immigration policy, 7–8, 62–63, 165
incarceration, of marginalized groups, 96, 142, 144–46, 149. *See also* carceral technologies
inclusion. *See* diversity, equity, and inclusion
Indigenous/Native people, 137–38, 150, 153–54
individualization, 34, 82, 90, 92–93, 110, 150
Industrial Revolution, Fourth, 44–46
industry capture, 64
inequities, intersecting, 84–85
information and communications technology, 36, 46
in loco parentis doctrine, 130–31
Institutional Review Boards (IRBs), 141, 142–43, 166
intelligence agencies, 14, 135, 137–40, 141
intelligence (IQ) testing, 34, 53
International Association of Campus Law Enforcement Administrators, 135–36
international students, 8, 62–63, 122, 133
internet, 1–2, 27, 106, 112, 117, 138, 152; of Things, 44
Internet2, 124
intruder detection technology, 119, 125
Iowa State University, 120
i-Pro, 119
iris scanners, 123
Islamophobia, 131–32, 156
Israeli-Palestinian conflict, 135
Ivy League, 53

Jackson, Virginia, 42–43
Jackson State College, 131
Jansson, Åsa, 92–93
Jasanoff, Sheila, 2
Jenkins, Steve, 108–9

Jewish students, 49–50
Jim Crow era, 53
Johnson, Erik, 112–13
Johnson v. NCAA, 66–67

Kent State University, 131
Key Worldwide Foundation, 48
Khaled, Leila, 74
Kim, Sang-Hyun, 2
Kizilcec, Rene F., 77
Kleinman, Molly, 122
Konradi, Amanda, 126
Kraft, David P., 85–86
K-12 education, 23, 50, 114–15, 117
Kwet, Michael, 164

labor exploitation, 157; of students, 19, 66–67, 76, 105–6, 133–34, 155
land-grant colleges, 49, 137–38
large language models (LLMs), 154–55, 157
Lascoumes, Pierre, 10
Latine students, 51, 134
learning analytics, 70, 71, 75–76
learning enhancement tools, 9–10
learning management systems (LMS), 19, 69–70, 74, 75, 76, 162–63
Le Galès, Patrick, 10
Lehigh University, 126
license plates, automated readers, 39, 129
Lindabary, Jasmine R., 142
Linkletter, Ian, 111–13, 158
Lockheed Martin, 139
Lombroso, Cesare, 119
Lorde, Audre, *A Burst of Light*, 99
Los Angeles Police Department, 127, 131, 146, 148–49
low-income communities, 3, 147
low-income students, 13, 57, 83, 91; admissions and recruitment, 50, 51, 52, 54, 56, 58, 64; exam proctoring and, 112–13; grading system and, 114–15
Lupton, Deborah, 93

machine learning, 75, 113, 129, 142; algorithms, 54–55, 59–60, 61; in criminal legal system, 59–60, 119,

144–45; in exam proctoring, 111, 113; racial bias associated with, 59–61, 119, 142–43, 144; in recruitment, 54–55, 58, 59–61; in research, 142–43; in securitization, 59–60, 119, 129

majority-white institutions, 64

male supremacy, 137

Manyika, James, 45

marginalized/minoritized communities, 160–61, 165–66; AI research and, 157; discriminatory personalization, 29–32; displacement by universities, 128–29; privacy rights violations, 7–8; securitization and policing practices toward, 69, 142–43, 145, 147, 149

marginalized/minoritized students: digital surveillance, 65; FRT and, 117–18; grading systems and, 115–16, 118; health services access, 91; mental health and wellness, 85, 91, 95–99, 109, 110; mental health and wellness self-tracking apps; recruitment, 56–57, 60; support services, 60–61

Mattelart, Armand, *The Information Society*, 26

McGee, Ebony O., 97

McKinsey & Company, 45

mental health and wellness self-tracking apps, 3, 18–19, 22, 82, 83–110; Amazon Echo Dots, 18–19, 89, 103–7; corporate use of data, 94–95, 108–9; dataveillance practices, 99–103, 109; implications for marginalized students, 85, 91, 95–99, 109, 110; mental hygiene movement and, 18, 86–89, 95; neoliberalism and, 90, 91–95, 99, 110; structural factors and, 19, 84–85, 90, 91–92, 94, 95–99, 109. *See also* WellTrack self-tracking app

mentorship, 117, 150

meritocracy, 114, 115

Mertz, Emily, 90–91

Meyerian psychobiology, 88–89

Meyeroff, Eli, 169

Miami University, 112–13

Microsoft, 13, 37, 153, 165

Middle Eastern and North African students, 62–63, 131–32

militarism/militarization, 14, 19, 39, 132, 133–34, 136, 137–38, 149, 165

Miller, Peter, 13

Mitchell, Nick, 15, 134, 137, 169

Montana State University, 83–84

mood mapping, 93–94

Moore, Phoebe V., 94–95

Morell, Michael, 139–40

Morrill Acts, 49

Moten, Fred and Stefano Harney, *The Undercommons . . .*, 169

multiversities, 44

Napolitano, Janet, 133, 134

National Collegiate Athletic Association, 66–67

National Conference of State Legislatures, 32

National Guard, 131

National Housing Act, 30–31

National Science Foundation, 139, 140

National Security Agency, 7

national security and defense, 7, 13, 19, 62, 122, 125, 131–32, 137–40

National Security Commission on Artificial Intelligence, 140

National Security Entry-Exit Registration System, 62, 122

natural language processing, 75

neoliberalism, 4, 5, 10–11, 13, 57, 68, 90, 91–95, 99, 125

Nestor AI, 71

Netflix, 3, 12, 28, 33, 59

neurodivergent students, 117

neutrality issue, in smart technology, 10, 31, 36, 42–44, 60, 68, 117, 128, 145, 146, 152

New America, 74

Newcastle University, 166–67

New Deal for Higher Education, 167–68

Newfield, Christopher, 15–16; *The Great Mistake . . .*, 16

New York University, 74

Next Generation Learning Challenges program, 75

Niemtus, Zofia, 38
Nixon, Richard M., 7
noncitizen students, 62–63, 64, 74
nontraditional students, 24
Northeastern University, 44–45, 104
Northern Arizona University, 23
N-Powered, 104, 105
nudges, 18, 22–23, 33, 34, 54, 59, 69; for retention, 70, 72, 73–74, 79; for self-care, 105
Nyaupane, Pratik, 151

Ochigame, Rodrigo, 68
OECD countries, 23–24
Office of Scientific Research and Development, 138
Ohio University, 83–84
Okechukwu, Amaka, 52
online learning, 4, 75–76, 118
OpenAI, 153–54, 155
open-source higher education communities, 159
Our Data Bodies, 166

Palmer, Iris, 74, 151
Pangrazio, Luciana, 152
Pardo-Guerra, Juan, 159–60
PATRIOT Act, 8
Pell Grants, 57
people of color, 7–8, 49–50, 137–38
personal information: aggregated and anonymized, 100–101; government access, 7–8, 63, 106–7; leaked or hacked, 123–24; legal protection, 106–7; rights to, 104; use in marketing, 100–102
personalization, 12–13, 16–17, 25–33, 47, 57–58, 119–20; development, 25–27; discrimination associated with, 29–32; e-commerce and, 16–17, 28, 33–36
plagiarism detection tools, 153, 162
Polaroid, 165
police/policing, 165; access to student data, 7, 107; actuarial methods, 68; campus police, 8, 19, 38–39, 127–36, 149; corporate partnerships, 70, 165; divestment movement, 149; FRT use, 119, 121–22, 142–43; international activities, 135; militarization, 19, 39, 132, 133–34, 136, 149; police violence, 96, 121–22, 127, 131, 134, 147–48, 166; racialized practices, 21, 39–40, 145, 147–48, 166; residential surveillance, 42; university partnerships, 127–29, 131, 146, 149. *See also* predictive policing
Ponsiglione, Chris, 123
predictive analytics, 3–4, 22–23, 25, 28, 35, 81; ethnographic study of, 43, 79; personalization based on, 25, 28, 33–34; in predictive policing, 13–14, 20–21, 39–40, 68, 69, 137, 140, 144–49, 160, 166; racial / social inequalities associated with, 13–14, 17, 20–21, 35, 39–40, 43, 56, 69, 79–81, 137, 140, 144–49, 160, 166; in recruitment, 17, 35–36, 56–57, 58; in retention, 17–18, 22–23, 35–36, 43, 72–73
predictive policing: corporate and government funding, 146; for financial crime, 160; opposition, 20–21, 148–49, 166–67; racialized bias and applications, 14, 39, 68, 146–49, 160; risk classification and management for, 68; university research in, 13–14, 20–21, 39–40, 69, 137, 140, 144–49, 166–67
PredPol, 39, 146, 147, 148–49
Princeton University, 86
Prinsloo, Paul, 164
privacy issues: vendors, 9, 76–77, 101, 124, 151, 153–54
privacy rights/issues, 3, 6–9, 17, 42, 105–8; Alexa-enabled devices and, 104; ChatGPT, 156–57; computer privacy settings, 111–12; dataveillance practices and, 99–103; FRT, 142–44; IRBs, 142–43; in mental health and wellness tracking, 99–103, 104; personalization, 29; predictive policing, 148; privacy laws, 151–52; reasonable expectation test, 143–44; research-related, 142–44; smart cities, 37, 40; Student Privacy Bill of Rights, 151–52; student records, 6–7
private equity firms, 76

privatization, 10, 15, 16, 17, 36, 40, 64; AI tools and, 160; cloud technology and, 75; of exam proctoring, 116–17; retention management and, 72–73, 74–76; in smart cities, 36, 40; student mental health implications, 82, 110
Proctorio, 20–21, 111–13, 114, 116, 117, 118, 158, 166
Proctortrack, 118–19
ProctorU, 112
profiling: criminal, 140; of prospective students, 55, 56–57; racial, 39–40, 59–60, 127–28, 141–42, 144–49
Project Maven, 165
public funding, for higher education, 2–3, 10–11, 16, 45, 50–51, 98–99, 136, 138, 167, 168; performance-based, 17–18. *See also* austerity politics
public land: corporate takeover, 41; university takeover, 128–29
punch cards, 5–6
Purdue University, 72, 83–84

racial integration, 130–31
racism/racial inequities, 3–64, 49–50, 61, 77–78, 79, 159, 160; admissions and recruitment, 48, 49, 51, 56, 59, 61, 63–64, 116; Black athletes, 66–67; campus policing, 127–28; ChatGPT, 156; discriminatory admissions/ recruitment policies, 49–50, 51–52, 53–54, 56–57, 61; FRT technology, 71–72, 118–19, 121–22, 142–43; grading systems, 114–18; K–12 education, 50; mental health tracking apps, 95–99; mismatch theory, 97; New Jim Code, 147–48; racial profiling, 39–40, 59–60, 96–97, 127–28, 141–42, 144–49; securitization, 113–14, 126, 128–29, 145, 149; student retention, 18, 72; surveillance-based, 29–31; techno-scientific, 137; university AI research, 14, 19, 137–38, 141–42, 144–49
radiofrequency analytics, 1, 163
Reagan, Ronald, 10, 50–51

recruitment, 3, 8, 11, 12–13, 17, 22, 47, 48–64, 119–21; budgets / financial issues, 57–59, 61; diversification, 52–53; equity-based, 57; financial aid leveraging strategies, 57–58; inter-institutional competition and, 54, 120–21; international students, 62–63; racial and economic disparities, 17, 49, 56, 59, 61, 63–64; social media posts surveillance in, 55, 61–63; student-recruitment officer interactions, 61
remedial education, 72–73
research. *See* university research
retention, 11, 64, 65–82, 119–20; at-risk students, 64, 67, 69, 73–74; digital tools, 70–76; discrimination and exploitation issues, 67, 76, 79–81; predictive analytics in, 17–18, 22–23, 35–36, 43, 72–73; privacy issues, 67, 74, 76–78; privatization, 72–73, 74–76; risk mitigation in, 65–82; student mental health and, 86, 108
Roberts, Dorothy E., 96, 148
Robertson, Cerianne, 36–37
Rodríguez, Dylan, 136
Roosevelt, Franklin D., 138
Rose, Nikolas S., 13
Rosenfeld, Hannah, 122
rural students, recruitment, 56
Russell Sage Foundation, 7
Ryerson University, 83–84, 91

Sadowski, Jathan, 42
safety. *See* securitization
St. Louis University, 103–5
Saravanan, Raja, 119–20
Savio, Mario, 5–6
scholarships, 51–52, 57–58, 115, 139
Scholastic Aptitude Test (SAT), 48, 53–55, 116
Schulte, Stephanie Ricker, 28
Schwab, Klaus, 45
Schwartz-Weinstein, Zach, 169
science and technology studies, 5
seamlessness concept, 23, 37–38, 119–20, 151

Seamster, Louise, 117–18
securitization, 1, 19, 39, 111–49; campus and student security, 8, 39, 103, 125–36; campus police's role, 38–39, 127–36; of grading system, 113, 114–18, 134; ideology of, 113–14; racialized, 126, 128–29, 145; of research, 137–49; safety apps, 113–14
segregation, 56, 87, 142
Selwyn, Neil, 5
September 11, 2001 terrorist attacks, 62, 122, 131–32
Servicemen's Readjustment Act (GI Bill), 50
sexism, 79, 156
sexual minority / gender-nonconforming / queer community, 32, 83
sexual violence/harassment, 125–27, 165
Shaffer, Christopher M., 12–13, 120
Siemens, 37, 40
Silicon Valley, 34, 73, 133, 139, 153
Simon, Caroline, 108
Siri, 59
SkyCop, 146
slavery, 67, 68, 96, 98, 137, 141–42, 148
smart cities, 17, 25, 36–47, 145–46; case examples, 36–37, 41; seamlessness concept, 23, 37–38
smart universities, 1–2, 22–47; coalition-based movement against, 150–69, 160–69; differentiated from smart campuses, 24–25; historical context, 5–7; proponents, 2, 3–4, 17, 19–20, 23, 24–25, 33–35, 37–39, 41–46, 73, 81, 91, 134–35, 146, 150; public policy and, 10–12; stakeholders' input, 41–42; students' attitudes toward, 150–51; students' decision-making power over, 158–59; students' expectations about, 4, 150–51
social control, 10, 86, 90, 130–31, 142
social inequalities. *See* economic and social inequalities
socialization: big data, 13–14, 20, 61–62, 84, 89; into digital economy, 150; mental health-based, 86–87

social media, 7, 8, 150, 167; surveillance of posts, 55, 61–63, 65–66, 129
Social Sentinel, 129
speech recognition technology, 153–54
SpotterEDU, 70–71
spyware, 111–14
standardized tests and testing, 52–53, 54; falsified scores, 48. *See also* Scholastic Aptitude Test (SAT)
Stanford University, 48, 141, 164
statistics, 26–27
STEM education, 139
stereotyping, 66, 74, 147–48, 156
Stovall, David, 97
Strike Debt, 167
strikes/strikebreaking, 19, 107, 133
structural inequalities, 2, 3, 4, 5, 43, 44, 46, 49, 64, 79, 80, 82, 150, 160–61; coalition-based movement against, 20–21, 160–69; grading system and, 114–17; mental health / wellness and, 84–85, 90; predictive policing and, 146; privacy rights and, 143–44, 151–52; recruitment process and, 63–64; smart city initiatives and, 36–37. *See also* racism / racial inequities
student activism, 50; 1920s, 86–87; among students of color, 51; critical digital literacy-based, 157–58; against digital exam proctoring, 20–21, 112–13, 158; Free Speech Movement, 5–6, 10; Occupy movement, 125, 132, 136; against smart university initiatives, 4–5, 164, 166; surveillance and suppression, 107, 122, 125, 129, 131, 132–34, 149
Student and Exchange Visitor Information System, 122
student debt, 16, 21, 50–51, 57, 98–99, 139, 160, 167
student-faculty ratios, 72–73, 82
Student Load Debt Center, 167
student loans, 32–33, 50, 167
Students for Sensible Drug Policy, 122
students of color, 13, 21, 31, 50, 51, 52, 54, 64, 72, 91, 97, 98, 115, 117, 166
Sunstein, Cass, 33

sustainability, 2, 36, 37, 38
Swauger, Shea, 118
Sweeney, Latanya, 31
Swift, Sharon, 35

tax-exempt status, of universities, 139
technocentrism, 34–35
technocratic management/power, 25, 42, 71, 162
technological solutionism, 73
technology-transfer movement, 139
techno-utopianism, 12
Temple University, 69
third-party doctrine, 106–7
Thirteenth Amendment, 67
Thomas, Angie, *The Hate U Give*, 156
Thomas Bravo (private equity firm), 76
Thornhill, Ted, 61
transfer process, 78
transparency, 7, 68, 82, 151, 153; in admissions, 55; of ChatGPT, 156–57; of exam proctoring software, 112–13, 116; national security implications, 138–39; in policing, 39, 148–49; privacy policies for, 9, 76–77; vendors and, 76–77, 151, 153–54
Transportation Security Administration, 118–19
Truman State University, 104
Trump, Donald, 7
tuition, 4, 10–11, 35, 50, 57, 58–59, 62, 78, 98, 104, 125, 132, 133, 167
Tulumello, Simone, 146–47
Turnitin, 153

UBTOS USA Inc., 127
undocumented students, 133–34
University of Alberta, 83–84, 90–91
University of Arizona, 132
University of British Columbia, 113
University of California, 76; Berkeley, 5–6, 10, 50, 132–33; Davis, 125, 132; Los Angeles, 48, 131, 146, 148–49, 166; Santa Cruz, 19, 63, 83–84, 115–16, 133–34
University of Colorado: Boulder, 113; Colorado Springs, 141, 143–44

University of Georgia, 123
University of Illinois Urbana-Champaign, 113
University of Lille, 36
University of Memphis, 13–14
University of Miami, 83–84, 132–33
University of Minnesota, 113
University of Nevada, Reno, 128
University of North Carolina, 70–71, 129
University of Southern California, 48, 123
University of Texas: Austin, 38, 48; Dallas, 104
University of Wisconsin, 120
university research, 4, 16, 137–49; AI technologies, 140–49; carceral technology, 13–14, 20–21, 39–40, 137, 138–40, 144–49, 166–67; corporate and government partnerships, 13–14, 19, 23, 39–40, 45, 138–41, 144; design justice approach, 165–66, 169; early techno-scientific racism, 137–38; national security, 138–41, 142–43; racialized applications, 39–40, 137–39, 141–42, 144–49
upgrading, 115
US Customs and Border Protection, 63
US Department of Defense, 127, 138–39; ARPANET and DARPA agencies, 138
US Department of Education, Office for Civil Rights, 52
US Department of Homeland Security, 129, 133
US Department of Justice, 29, 48; Civil Rights Division, 52
US Department of State, 63, 138–39
US Immigration and Customs Enforcement (ICE), 70, 122, 132, 136, 165
US Supreme Court, 51, 167; affirmative action decision, 51–52; *Brown v. Board of Education*, 130–31; *Carpenter v. US*, 106–7; *Katz v. US*, 143–44; *Regents of the University of California v. Bakke*, 51

Vanderslot, Jodie, 91
Varian, Hal, 29–30
Varin, Bryan, 123

vendors, of smart technology, 2, 9, 16, 40, 120–21, 124, 128; privacy and transparency issues, 9, 76–77, 101, 124, 151, 153–54
venture capital, 78
Verificient, 112
video analytics technologies, 38–39, 113–14
Virginia State University, 13
visas, 8, 62–63, 115, 122, 133

Washington University, St. Louis, 63
Watkins, Grace, 135
Watters, Audrey, 5, 34
Weagle, Stephanie, 38–39, 40
wealth inequality, 56, 98, 150
web browsing, 22, 26, 55, 111–13
websites, of colleges and universities, 55, 56
WellTrack self-care tracking app, 18–19, 83–85, 89–92, 93–103, 104, 107–10; Activity Scheduler tool, 98–99; CBT-based, 18, 83, 92–95; Cognitive Distortions Quiz, 95–97; Mood Check / Heatmap, 93–94, 95; privacy policy, 99–103; structural racism and, 95–99; Thought Diary, 95
West Virginia University, 168
Whisper AI, 153–54
White, Hayden, 131
white collar crime, 160
white privilege, 29, 54
white students, 51–52, 56–57, 121–22, 126
white supremacy, 116, 137–38, 148
Whitman, Madisson, 43, 79
Whittaker, Meredith, 140
Wilder, Craig Steven, 137
Williams, Frankwood, 86–88
Williamson, Ben, 5
Williamson, Kimberly, 77
Winfield, Ann Gibson, 114
W.K. Kellogg Foundation, 124
women's colleges, 49–50, 130
Wondergem, Taylor, 134
Wood, Benjamin D., 34
workforce: activism, 70, 164–65; of digital economy, 13; discrimination in, 164–65; educational preparation, 23–24; emotion recognition surveillance, 121; Fourth Industrial Revolution and, 44–46; income inequality, 64; mental wellness self-tracking, 94–95, 108–9; workplace surveillance, 61–62
World Bank, 36
World Economic Forum, 45–46

Yale University, 48, 130, 165
Yip, Christine, 35
York University, 83–84, 91

Zegart, Amy, 139–40
Zen Room app, 102–3
Zhang, Yuchen, 35
ZIP codes, 56
Zoom, 74, 133–34

Explore other books from HOPKINS PRESS

UNIVERSITIES ON FIRE
Higher Education in the Climate Crisis
BRYAN ALEXANDER

"A compelling account of the intersection of academia and climate change from one of our foremost futurists."

—Michael M. Crow,
coauthor of *The Fifth Wave: The Evolution of American Higher Education*

HOW COLLEGES USE DATA
Jonathan S. Gagliardi

"A comprehensive playbook for the creation and ethical adoption of a scalable culture of data that focuses on defining evidence-based and equity-driven aspirations."

—Nancy Zimpher,
Chancellor Emeritus,
State University of New York

Unsettling the University
CONFRONTING THE COLONIAL FOUNDATIONS OF US HIGHER EDUCATION
Sharon Stein

"A brilliant, decolonial critique of how liberal, inclusion-focused, knowledge-focused approaches tend to reproduce colonial patterns, this book presents alternative approaches to justice in higher education."

—Eli Meyerhoff, Duke University

JOHNS HOPKINS UNIVERSITY PRESS | PRESS.JHU.EDU |